PENGUIN BOOKS

THE GREEN CONSUMER SUPERMARKET GUIDE

JOEL MAKOWER is a Washington, D.C.-based writer and lecturer on consumer, environmental, and business topics, and editor of *The Green Consumer Letter*, a monthly newsletter. He is coauthor (along with John Elkington and Julia Hailes) of *The Green Consumer* and *Going Green: A Kid's Handbook to Saving the Planet*. He is a graduate in journalism from the University of California at Berkeley. A former consumer columnist, he is author or co-author of more than a dozen books.

JOHN ELKINGTON is one of Europe's leading authorities on the role of industry in sustainable development. He runs an independent consultancy whose clients have included British Petroleum, the Nature Conservancy Council, and the United Nations Environment Programme. Having written numerous books and reports, he is also the co-author, with Tom Burke, of *The Green Capitalists*.

JULIA HAILES has worked in advertising and TV production. In 1987, she helped to set up SustainAbility, "the green growth company," with John Elkington and Tom Burke. The company aims to promote environmentally sustainable economic growth.

THE GREEN CONSUMER SUPERMARKET GUIDE

Joel Makower

with John Elkington and Julia Hailes

A Tilden Press Book

PENGUIN BOOKS

PENGUIN BOOKS
Published by the Penguin Group
Viking Penguin, a division of Penguin Books USA Inc.,
375 Hudson Street, New York, New York 10014, U.S.A.
Penguin Books Ltd, 27 Wrights Lane,
London W8 5TZ, England
Penguin Books Australia Ltd, Ringwood,
Victoria, Australia
Penguin Books Canada Ltd, 2801 John Street,
Markham, Ontario, Canada L3R 1B4
Penguin Books (N.Z.) Ltd, 182–190 Wairau Road,
Auckland 10, New Zealand

Penguin Books Ltd, Registered Offices:
Harmondsworth, Middlesex, England

First published in Penguin Books 1991

3 5 7 9 10 8 6 4 2

Published by arrangement with Victor Gollancz Ltd.

Grateful acknowledgment is made for permission to reprint
excerpts from *Silent Spring*, by Rachel L. Carson,
Copyright © 1962 by Rachel L. Carson. Reprinted by
permission of Houghton Mifflin Company

CIP data available

Printed in the United States of America

Contents

Nonfood Items

Part Three: The Green Supermarket

Part Four: Action Guide

Acknowledgments

A great many individuals and organizations helped in this effort, and they deserve recognition and thanks:

Jessica L. Weiss, research director, provided valuable assistance at every stage of this project. She was assisted ably by Carla Garrison, who oversaw the research on supermarkets; Mo Good; and Amy Gilman. The staff at the Council on Economic Priorities also provided valuable resources and was there with ready answers to our many questions, and we owe sincere gratitude to Alice Tepper Marlin, Jonathan Schorsch, and Robert Rubovitz.

Thanks also to: George Arndt, National Food Processors Association; Eric Auth, Jerry's IGA Market; Jill Beresford, Beresford Packaging; Linda Brown, Green Cross Certification Co.; Dr. Gerald Combs, Department of Agriculture; Dave Davis, Duro Bag; Norman Dean, Green Seal; Philip Dickey, Washington Toxics Coalition; Pamela Dorman, Penguin Books; Aileen Dullaghan, Food Marketing Institute; Jayme Kerr, American Society for Testing & Materials; William J. Klaisle, Stone Container Corporation; Janet Kraybill, Penguin Books; Peter Larkin, Food Marketing Institute; Lisa Lefferts, Center for Science in the Public Interest; Jatal Mannapperuma, U.C. Davis; Michael Marks, Stone Container Corporation; Wendy Noble, Food Marketing Institute; Leo Pederson, National Food Processors Association; Lars Peterson, Food Marketing Institute; Greg Phillips, Ashdun Industries; Stanley Rhodes, Green Cross Certification Co.; Randy J. Rosenberg; Cha Smith, Washington Toxics Coalition; Hugh Symons, American Frozen Food Institute; Gale Warnquist, Plastic Bottle Information Bureau; Ed Weary, Flexible Packaging Institute; Sue Wilkinson, Food Marketing Institute; and Jeanne Wirka, Environmental Action Foundation.

AUTHORS' NOTE

The authors and the publishers of the American edition of this work have prepared this work and set forth the views contained in it in good faith.

The omission of any particular product, company, or any other organization from this work implies neither censure nor recommendation. Neither the authors nor the publishers of this edition of the work, however, warrant effectiveness or performance of any product set out in this work, and the reader and consumer must exercise his or her own judgment when determining what criteria to apply when judging whether or not he or she should purchase any particular product from any of those companies or other organizations which are mentioned in this work. Any reader requiring further information concerning any particular product should write directly to the supplier or manufacturer of that product.

The authors and the publishers are willing to consider any matter which is held to be inaccurate and upon satisfactory factual and documentary confirmation of the correct position will use all reasonable endeavors to amend the text for the next reprint or edition.

While every care is used and good faith is exercised in assembling and publishing the material in this book, no warranty as to the properties or safety of any product is imputed by the inclusion of any such product in this work, and consumers should address themselves to the manufacturers of any product in respect of which they may wish to be informed as to its qualities.

Part One

T·H·E P·O·W·E·R OF G·R·E·E·N

INTRODUCTION

Let's start with the basics: Every time you open your wallet, you cast a vote "for" or "against" the environment.

This is more powerful than you might imagine. First and foremost, the marketplace—whether the supermarket, hardware store, or appliance showroom—is not a democracy. It doesn't take 51 percent of people "voting" in any one direction to affect environmental change. Far from it. In fact, a relative handful of shoppers can send shock waves through an industry simply by making good, "green" choices.

Here's an example. In April 1990, three major tuna canning companies announced that they had made a revolutionary decision: They would stop buying tuna caught in a way that harmed dolphins. In the past, millions of dolphins a year were being needlessly killed simply because they became caught in tuna fishermen's nets. When these three tuna companies made their announcement, they said nothing about "protecting dolphins," or even about "saving the earth." They spoke instead of "consumer pressure."

What's amazing about all this is that it was a relatively small number of consumers who were "pressuring" the tuna companies—probably less than a million active individuals, according to some reports. That's less than 1 percent of the American marketplace. In fact, during the twelve months preceding the tuna companies' announcement, *tuna sales had barely changed.* And so the "votes" of a very small number of individuals revolutionized an entire industry.

Another revolution took place a few months later. In October 1990, McDonald's Corporation announced that it was abandoning its polystyrene foam "clamshell" hamburger boxes in favor of paper products. In making their announcement, McDonald's officials didn't claim that there was anything wrong with its plastic packages. But "our customers just don't feel good about it," said the company's president. "So we're changing." And again, while McDonald's had been the subject of consumer protests over the amount of unrecyclable trash it created, *its sales had increased during the preceding year.*

There are other, less dramatic revolutions happening nearly

every day. Since Earth Day 1990, practically every consumer products company seems to have examined its products and policies through a "green" lens, taking a hard look at how sales might be affected by Americans' growing concern about the impact of their purchases on the environment. Across America, companies are listening to consumers' concerns for a cleaner environment, and watching carefully the way we "vote" with our dollars.

This book will show you how to cast votes "for" Planet Earth whenever you shop for food and groceries. It is intended to help you sort through the often confusing world of Green Consumerism to choose products that are environmentally sound.

Why should you do this? For starters, as you'll see, some of the products we buy contribute to environmental problems. You may be surprised at the ways this can happen. For example, the manufacture of a paper towel or napkin can contribute dangerous pollutants downstream from a paper mill. The use of some aerosol products contributes to urban smog. Many products are packaged excessively in materials that are neither recycled nor recyclable. Some products are made by companies that have poor environmental records. Buying products from companies that pollute supports their lack of concern for the environment.

Make no mistake: We're not suggesting we can simply shop our way to environmental health. Part of being a Green Consumer is learning when *not* to buy—when more is not necessarily better. (See "The Fourth 'R': Reduce," page 27.) By making the right choices, you can help to minimize the pollution and waste created by many "un-green" products and companies.

A Little History

Why is all of this happening now? The growing concern over the environment is certainly a factor, but that's not the whole story. Part of the answer has to do with the fact that as consumers, Americans are going through a transformation. We've graduated in a fashion from the process that social historian Daniel Yankelovich once called "Enlightened Consumerism," where we learned not to take as gospel everything we hear in advertisements or read on labels. Today, we're more than merely enlightened—we're empowered. When it comes to protecting our

How to Be a Green Consumer

When it comes to evaluating and choosing green products—those that cause minimal impact on the environment—here are ten basic guidelines you should consider when evaluating a product:

- ❑ Look for products packaged in readily recyclable materials such as cardboard, aluminum, or glass.
- ❑ Don't buy products that are excessively packaged or wrapped.
- ❑ Look for products in reusable containers, or for which concentrated refills may be purchased.
- ❑ Look for products made from the highest content possible of recycled paper, aluminum, glass, plastic, and other materials.
- ❑ Choose simple products containing the least amount of bleaches, dyes, and fragrances.
- ❑ Look beyond the products to the companies that make them. Support those with good environmental records.
- ❑ Don't confuse "green" with "healthy"; not everything packaged in recyclable packaging is necessarily good for you or the environment.
- ❑ Remember that there are few perfectly green products.
- ❑ Keep in mind that doing something is better than doing nothing.
- ❑ It doesn't matter much whether you carry your purchases home in a paper or plastic bag. What's important is that you reuse or recycle whatever bag you get.

pocketbooks and our bodies, we're much more willing than ever before to take matters into our own hands—to stand up and be counted, whether in a nationally organized boycott, a press for legislative or regulatory reform, or simply an additional piece of data we take with us on our next shopping trip.

From "Me" to "We." At the same time, as other observers have pointed out, we're moving gradually from the "me" generation into the "we" generation. Self-interest certainly hasn't disappeared, of course, but it's been tempered by the fact that after a decade or so of "feel-good" national leadership, many of us don't feel so good about the world around us. There are a great many problems—including the environment—that are getting worse. We've pretty much given up on government and industry's ability to solve the problems for us, and technological solutions always seem to be another frustrating couple of years away. As a result, all we have left to count on is ourselves. And fortunately, a great many people seem to be responding to today's challenges, from housing and homelessness to drug abuse and education.

The fact is, many problems have gotten more personal, more widespread, and much closer to home. That certainly is the case with the environment. Environmental problems are no longer limited to a few localities. It's no longer simply the sensational far-off event—the Love Canal hazardous waste leaks, a burning Cuyahoga River, a malfunctioning Three Mile Island, or even a gushing Exxon *Valdez*. Almost every community in the land is coming to grips with some kind of environment scourge, from overflowing landfills to dirty air and water.

And there are global problems that simply didn't exist a decade ago. Our rain can be poisonous, our summers may be hotter, we may risk skin cancer with our suntans. So there's a paradox: At the same time some environmental problems are getting closer to home, others seem abstract and far away.

Shopping 'Til We Drop. While all this has been going on, we've become rampant consumers. Nearly uninterrupted economic growth has helped to create a glut of goods in the marketplace. More than 20,000 new products land on supermarket shelves every year, a rate that does not appear to be slowing down. Those products arrive in stores in an astonishing assortment of materials, sizes, and formats. It's no longer simply "regular size" and "family size." From aerosols to zip tops, the possibilities have been limited only by financial and technological constraints. Plastic has helped create the packaging boom, but equally disturbing is packaging made of several materials—combinations of plastics, for example, or laminates consisting of paper, plastic, and foil. All of these present new problems and challenges when it comes time to recycle, reuse, or otherwise dispose of such trash. Ironically, some of these packaging options—aerosols and aseptic "juice boxes," for example—are being presented by their creators as environmentally superior.

It's been a colorful carnival of choices. But now we're finding that along with this apparent prosperity comes some tough problems, not the least of which is how to reconcile the resources being used and the trash created by it all. We've come to expect to pay extra for convenience, but we're just beginning to fully understand how high the price tag for convenience may really be.

Keeping Things in Perspective

It's important to keep things in perspective. Green Consumerism is not the ultimate solution to the world's environmental problems. The problems are simply too many and too complex to be resolved merely by smart shopping. For example, for all the debate over paper versus plastic shopping bags, or glass versus aluminum containers, the environmental differences between these choices are obliterated by the energy and pollution impact of driving just one mile and back to the supermarket.

That doesn't mean, however, that being a Green Consumer is not important. Once you begin to incorporate the environment into your everyday shopping trips, it won't be a major step to then incorporate it into other aspects of your life—such as voting, investing, and educating your kids. And so in that light, Green Consumerism takes on a new importance: If we can't sell the environment at Safeway and 7-Eleven, we probably won't be able to sell it in the schools, the statehouse, and the Senate.

That makes your everyday shopping trips the key to saving the earth.

Who's Shopping Green?

It seems that nearly every month there's another study that shows that Green Consumerism is a driving force in the American marketplace. One study shows that 89 percent of Americans would pick Product A over Product B if Product A was kinder and gentler to the planet. A Gallup Poll found that about three out of four Americans consider themselves "environmentalists." Another tells us that 82 percent of us are voluntarily recycling some materials. Still another survey revealed that three-fourths of all Americans are willing to pay a bit more for environmentally responsible products. A Louis Harris poll revealed that "a clean environment" ranked second, just behind "a happy family life" among American's priorities. ("A satisfying sex life" ranked only in fifth place.) Clearly, the eco-number-crunchers are working overtime.

You are advised to not accept these polls and surveys at face value. While there certainly is increased interest in the environmental impact of one's purchases, there is a very big step from environmental *concern* to environmental *consumerism*. For example, that eight out of ten Americans claim they are recycling "some materials" is a bit suspect in a nation that recycles only about a tenth of all its trash. The fact is, when it comes to the environment, people are more likely to give the "right" answer than the "real" answer. But even if the polls are off by a factor of three or four or five, we're still talking about a significant chunk of the marketplace.

The Price Is Right

One of the most infuriating questions asked by some of the polls and surveys is how much extra Americans would be willing to pay for green products. In most cases, the majority of respondents claim that they would be willing to pay between 5 percent and 15 percent extra for environmentally responsible products. But the question is itself misleading: It furthers the myth that green products are by their very nature more expensive than their "ungreen" counterparts.

The fact is this: Being a Green Consumer can be a money-saving proposition. And with good reason. At its core, Green Consumerism has to do with cutting waste and making the most of one's resources. You can't help but get more for your money in the process.

This notion isn't based on mere speculation. In 1990, two supermarket industry consultants asked Information Resources Inc., a market-survey firm, to track several green products in a major West Coast supermarket chain. The products included those made from or packaged in recycled material, photo- and biodegradable products, those whose "use does not pollute water or air," and products "made using a cleaner process." The researchers were surprised by their own findings. "Consumers weren't in fact paying more for these brands," they told *Progressive Grocer*, a trade magazine. "On average they actually paid be-

Cold Comfort

In this book, you'll spend a lot of time considering what goes in your refrigerator. But you'd be wise to take a few minutes to consider the refrigerator itself.

You may be surprised to learn that your refrigerator may be the single biggest electricity user in your home. In fact, cumulatively, America's refrigerators use the output of about 25 large power plants—about 7 percent of the nation's total electricity consumption and more than half the power generated by the nation's nuclear power plants.

Refrigerators need not be so energy-hungry. Models now on the market use half the energy of their forebears in the 1970s. According to one estimate, if every household in the United States had the most energy-efficient refrigerators currently on the market, the energy savings would eliminate the need for about ten large power plants.

But there is still room for improvement. Congress has passed legislation to require refrigerators and freezers built after 1992 to be 25 percent more efficient than they were in 1990. And Lawrence Berkeley Laboratory says that they could be as much as 35 percent more efficient, and still be competitively priced.

The investment in a new refrigerator may be well worthwhile. Keep in mind that the purchase price of any appliance is by far the smallest cost involved in owning it. The cost of the electricity to run a refrigerator for 15 to 20 years can be more than three times the appliance's purchase price. So, the extra money you spend to buy an energy-efficient refrigerator will be more than repaid over the years through reduced electric bills. And as energy costs continue to rise — the average residential electricity rate has more than tripled during the past 15 years and is certain to continue to rise — the investment in energy-efficient appliances will yield even bigger dividends.

tween 8 percent and 22 percent less for the green brands contrasted to the average prices paid in the category." But this isn't universally true: Some greener products do cost considerably more than their less-green counterparts.

So, being a Green Consumer can be both ecological and economical.

The Many Shades of Green

One of the first things you find when you take a look at the world of Green Consumerism is that there are few *perfectly* green products. In fact, there are many shades of green. There are green products packaged in un-green packaging, green packaging enclosing un-green products, and green products made by decidedly un-green companies. To make matters worse, you can buy green products from un-green supermarkets, and un-green products at green supermarkets.

It's also important to point out that compiling a list of products that are safe for the environment is simple: There are none. That's right: Nothing is safe for the environment. Its mere existence means it's had some environmental impact. The goal, then, is to choose products that have a minimal impact on the environment. (One of the first things you can do immediately is to discount any product whose manufacturer claims it to be "safe for the environment.")

Confused? That's understandable. But don't let it discourage you. All it takes to begin making green choices are a few ground rules—and this book.

What Makes a Product Green? There are few standards or laws that determine what makes a product environmentally responsible. But there are some basic criteria we use when we look at a product's greenness:

❑ A green product is one that is not dangerous to the health of people or animals.
❑ It causes minimal damage to the environment during its manufacture, use, and disposal.
❑ It does not consume a disproportionate amount of energy or other resources during its manufacture, use, or disposal.
❑ It does not cause unnecessary waste, due either to excessive packaging or to a short useful life.
❑ It does not cause unnecessary cruelty to animals.
❑ It does not use materials derived from threatened species.
❑ Ideally, it should not cost any more that its "ungreen" counterpart.

That's a tall order, to be sure. And few, if any, products contained in this book meet all of these criteria. But there is no question that due to increased consumer interest, a growing number of products are coming closer to hitting the mark.

Don't be discouraged by this lack of perfection. There are still green choices to be made in nearly every supermarket aisle. Many familiar products may be deemed "green" simply because they have minimal packaging, or because they are packaged in recyclable containers. The point is, there are plenty of green products to choose from, if you know what you're looking for.

The E-Factor. You already have a set of standards and preferences you bring with you when you walk the supermarket aisles. You think about price, of course, and quality. You consider manufacturer reputation and availability. And you think a lot about your own personal taste.

Each of those factors plays a role differently with every purchase. When choosing a brand of chicken soup, for example, you might say to yourself, "They all taste the same to me, so I'm going to pick the cheapest one." In that case, price is the determining factor. Or you might say, "I've always found Smith's to be a quality brand, so no matter what it costs, I'll buy their chicken soup." And so, brand name and reputation become the deciding factors. We make these decisions—consciously or unconsciously—with nearly everything we buy. These are the tools with which we shop.

Being a Green Consumer means adding another tool: the E-Factor, the environment. So, in choosing your brand of chicken soup you might decide, "I'll choose this one because it has fewer layers of packaging." Or, "This company has a good environmental record, so I'll choose their brand." The E-Factor may not play a role in every purchase, but neither does price, brand name, or quality. But you may be surprised to find how in many purchases the E-Factor does play a role.

The more you consider the E-Factor, the more you will find it a useful way to look at products—not just in the supermarket, but also in the drug store, the hardware store, the appliance center, the garden center, the automobile showroom—just about anywhere else you shop.

A Moving Target? The standards of what is "green" are changing constantly. Almost every week there is a new finding or a new technological breakthrough that raises questions or provides new answers on the nature of green products. At one point, for example, biodegradable and photodegradable trash bags and diapers were thought to be an acceptable solution to solid waste problems. It then became apparent that manufacturer claims of degradability weren't really true, and that in some ways, degradable products were *worse* for the environment than nondegradable ones because when they degrade, they break up into plastic powder that can pollute water.

There have been other reversals. At the beginning of 1990, for example, it was not technologically possible to manufacture paper shopping bags out of recycled materials, and plastic shopping bags could not themselves be recycled. By the end of that same year, both had changed: paper bags made from 100 percent recycled material were being manufactured in large quantities, and several national grocery chains had set up plastic bag recycling programs.

Keeping abreast of all this can be challenging. As one controversy is settled, another seems to sprout up in its place. But there are many solid facts about Green Consumerism—the complete and unlimited recyclability of aluminum and glass, for example—that will not change. And there is much you can do even today's information to be a Green Consumer.

How Our Purchases Affect the Environment

It is much easier to make good, green choices if you understand the environmental impact of some of the things you buy. We'll delve more deeply into some of the more specific problems in the Supermarket Checklist in Part Two of this book. For now, here is an overview of the leading environmental problems and how they may be affected by our purchasing decisions.

Air Pollution

❑ **Cars.** There's little question that the number-one cause of air pollution are automobiles. So what you drive and how you drive are important choices you can make. Consider your drive to the supermarket. It is probably located within a few miles of your home. That short driving trip may not seem particularly wasteful, but short trips in busy traffic are far more polluting than highway driving. It involves more stop-and-go driving, which decreases fuel economy. If your car is cold—if it hasn't been driven in a few hours—it will use more gas than when it has warmed up. The first five miles of driving are said to be the least fuel-efficient; after that the car warms up and operates more efficiently.

❑ **Filling the Tank.** Even the act of filling your car with gasoline causes serious pollution. Among the vapors emitted during filling is benzene, a chemical that causes leukemia and blood disorders. About 80 percent of airborne benzene is believed to be released at gas pumps as vehicles are being filled, or later when they are running. When you "top off" your tank—that is, fill it as far to the top as possible—you cause additional problems: When the car warms up, the gas becomes heated and expands. Some of the gas spills out, dumping even more fumes into the atmosphere.

❑ **Transporting Food.** Even if you walk or take public transportation to the supermarket, you still contribute to vehicle exhaust. How? By purchasing products shipped great distances to your local stores. According to the Worldwatch Institute, the average mouthful of food travels about 1,300 miles from farm field to dinner plate. Many products, especially produce, are shipped from overseas. For example, one-fourth of the grapes eaten in the United States are grown 7,000 miles away in Chile. This happens even though many of these products are grown or produced locally. Seeking products and produce from local sources not only reduces pollution by reducing transportation needs, it also helps your area's economy.

❑ **Energy Production.** Cars aren't the only problem. Electric-generating plants are another major source of air pollution when they burn coal, oil, or natural gas. The emissions of power plants, as you'll see, also contribute to other environmental problems. By purchasing energy-efficient items, from appliances to light bulbs, you can help reduce energy consumption. During certain times of the day, that enables utility companies to use their most energy-efficient, least-polluting generators. It also helps to reduce the need to build additional power plants.

❑ **Polluting Products.** Some products you buy directly pollute the air. Aerosols—including hair sprays, deodorants, cleaners, and paints—contain ingredients that all contribute to urban smog. (For more about aerosols, see "The Problems of Packaging," page 39.) Barbecue charcoal lighter fluids also spew pollutants into the air. It's no coincidence that Los Angeles, one of the most polluted cities in the United States, has created a comprehensive plan that includes banning many aerosols and all charcoal lighter fluids in coming years.

Water Pollution

❑ **Farming.** When we think of dirty water, we typically imagine huge industrial pipes discharging disgusting looking water into rivers and streams. Some industries still seriously pollute our water supply, but there are other sources. Agriculture is one big source. Modern agricultural methods use huge quantities of pesticides and fertilizers, millions of pounds of which reach groundwater or surface water before degrading. Federal investigators have detected 17 pesticides in groundwater in 23 states as a result of agricultural runoff. Waste water polluted with animal wastes also pollutes the water supply. The production of beef is particularly water-intensive. According to John Robbins, author of *Diet for a New America,* "To produce a single pound of meat takes an average of 2,500 gallons of water—as much as a typical family uses for all its combined household purposes in a month." Much of this water eventually drains off the farm and into surface or groundwater.

❑ **Manufacturing.** Making some of the things we buy contributes to water pollution. Paper manufacturing, for example, is a highly polluting business. Even making paper from recycled pulp is polluting, but less so than making paper directly from trees. Other big polluters include the agriculture, mining, and chemical industries.

❑ **Household Cleaners.** It's ironic, but some of the things we use to make our lives cleaner contribute to making our water dirtier. Our cities' water filtration systems do a pretty good job of flushing out the various chemicals we flush down the drain when we clean. But about 10 percent of the chemicals remain, and gets returned to our drinking supply. Other cleaning ingredients cause harm to marine animals and plants.

❑ **Garbage.** An amazing amount of litter and other trash ends up in rivers, lakes, and oceans, some of which harms fish and other animals. Plastics in particular cause the most harm, especially when they are mistaken for food and ingested by these creatures. A lot of the plastic found inside dead animals comes from packaging materials—plastic yokes used to bind six packs of soda and beer cans, bits of polystyrene from foam hamburger boxes, and plastic produce bags, to name just a few things.

Global Warming

❑ **Energy Production.** Turning fossil fuels into energy—whether in automobiles or electric-generating plants—creates carbon dioxide (CO_2), the number-one cause of global warming, also known as the "greenhouse" effect. About 60 percent of the world's electricity comes from burning fossil fuels. As CO_2 and other gases (principally chlorofluorocarbons, methane, and nitrous oxide) are released into the atmosphere, they keep the sun's heat from escaping the earth's surface, much like a greenhouse traps the sun's rays inside its glass panes. The result: The earth's surface temperature may increase a few degrees. Even a small rise in temperature could melt polar caps, thereby raising ocean levels a few feet; several major cities and miles of coastline could end up under water. The resulting weather changes could upset agriculture, making some growing areas arid, while making some other areas suddenly fertile. Conserving energy and purchasing energy-efficient products are two of the best solutions to avoid these serious and perplexing problems.

❑ **Automobiles.** Again, cars are a big culprit here. Although U.S. automobiles go about 50 percent further on a gallon of gas than they did in 1974—spewing 2,000 fewer pounds of CO_2 per car—there is still room for improvement. Improving new-car fuel economy to 40 miles per gallon could cut carbon emissions by 124 million tons—about 10 percent of current U.S. emissions. Such improvements would save money, too: According to the American Council for an Energy Efficient Economy, the 40 mpg standard would add about $500 to the price of a car, but would save $2,000 over the car's life in reduced fuel consumption.

❑ **Food Production.** Traditional farming methods use nitrogen-based fertilizers, which have been found to contribute methane gas into the atmosphere; methane is one of the principal gases contributing to global warming. Another source of methane is livestock production, which, together with rice paddies and the breakdown of waste in landfills, contributes nearly a fifth of global warming gases.

Acid Rain

❑ **Energy Production.** Besides contributing to air pollution and global warming, electric utilities also play a principal role in the creation of acid rain. As utilities burn high-sulfur coal, the airborne emissions mix with water vapors and other chemicals to create acids. These, in turn, fall to the ground, not just as rain, but also as sleet, snow, hail, and fog. The acids can peel paint from cars, kill marine life in rivers and lakes, and kill trees and other vegetation. Even more important, acid rain is being seen increasingly as a threat to human health. The chemicals in acid rain can harm the respiratory system, with the very young and the very old being at greatest risk. A congressional study blames acid rain for contributing to some 50,000 premature deaths a year in the U.S. and Canada.

❑ **Automobiles.** Acid rain is still another environmental legacy of our automobile society. Nitrogen oxides in automobile and truck exhaust are another principal ingredient in the creation of acid rain. There is little argument among the experts that reducing automobile use in acid rain-plagued parts of the country will lead to a significant reduction of that problem.

Ozone Depletion

❑ **Manufacturing Processes.** The principal cause of depletion of the ozone layer—a thin layer of atmosphere between 12 and 30 miles above the earth's surface that shields the earth from most of the sun's harmful radiation—are chlorofluorocarbons (CFCs) and halons. These two chemicals are used in a wide range of products, from plastic foam to printed circuit boards to portable fire extinguishers. When CFC molecules escape into the atmosphere during the products' manufacture, use, or disposal, they destroy molecules of ozone; each CFC molecule can destroy up to 100,000 ozone molecules. It takes CFCs between 15 and 25 years to rise into the sky before it reaches the ozone layer, and each molecule takes about 150 years to break down, so today's CFC emissions will have a long and destructive legacy. Halons, while not as widely used, have up to 10 times the ozone-destroying power of CFCs.

❑ **Car Air Conditioners.** CFCs (also known as Freon) are the principal refrigerant used in the 90 million cars and light trucks in the U.S. with air conditioning, representing about 25 percent of CFC use. While the Freon in air conditioners is contained in a closed system, the system can leak or break, spewing CFCs into the atmosphere. CFCs also escape when cars are junked or destroyed. A lot of the problem can be solved by regular maintenance of car air-conditioning systems so that leaks can be found and repaired early on. Some mechanics use devices that safely capture the coolant during repairs, then recycle it for reuse, making the process even less damaging to the environment.

❑ **Appliances.** Cars aren't the only appliances that use CFCs. Home air conditioners as well as refrigerators and freezers also use Freon. Again, when they leak, or when they are repaired, Freon can leak into the atmosphere. As with enlightened auto mechanics, several appliance service companies have devices that safely capture and recycle Freon during appliance repair. In addition to their use as a coolant in refrigerators, CFCs are also contained in the foam insulation inside the appliances' walls and doors. When disposed of, the foam deteriorates and the CFCs escape.

Vanishing Species

❑ **Products from Endangered Rain Forests.** Some of the most valuable parcels of real estate are being destroyed at alarming rates. Most of these are overseas, in the rain forests of tropical regions. These dense, lush environments contain millions of plant, insect, and animal species that may be of use as food, fuel, and medicine. For example, a plant called the rosy periwinkle has been found to cure some cancers. The plant is native to parts of Madagascar, where 90 percent of the forested area has been destroyed. Thousands of tropical plants have edible parts, and some are extremely nutritious and easy to grow. Other trees are useful as everything from dental cures to diesel fuel. When these forests are destroyed—usually for cattle raising or to harvest hardwoods—it destroys some species forever. Compounding the problem is that when trees are burned to clear fields for farming, carbon dioxide is released into the atmosphere; the burning of tropical forests contributes about a fifth of the global warming problem. Few of the products that come from these rain forests reach the U.S., but some of them do, including hardwoods used for everything from disposable chopsticks to expensive furniture.

❑ **Products from Endangered Species.** Some of these products are made from ivory, for which between 200 and 300 African elephants a day are killed. Although ivory imports have been banned by the U.S. and many other countries, a great deal of it still finds its way across American borders, in the form of tourist trinkets from the Far East. Animal furs are another product that encourages the killing of endangered species. Even *faux* furs from synthetic materials are considered unacceptable because they encourage the wearing of animals. Besides, most of these synthetic materials are made from petroleum-based products, which are nonrenewable resources and which contribute pollution and global warming during the manufacture process.

Solid Waste

❑ **Packaging.** The fact that we throw away a lot of wasteful trash is widely known, and packaging has become a symbol of that wastefulness. Packaging represents just under a third of all household trash—some 56.8 million tons a year nationwide. Much of this material can be readily recycled, and some of it can be recycled with a bit more effort, but a great deal of packaging is destined to end up in landfills. While packaging serves a purpose—it provides sanitation, convenience, and space for product information—American manufacturers have outdone themselves in recent years trying to attract shelf space and consumer attention in the crowded supermarket aisles. Of greatest concern are multi-materials—packages made from layers of plastics, aluminum, paper, and other materials, making recycling efforts all but impossible. But it isn't just the amount of waste ending up in landfills that causes problems. Some types of packaging contain chemicals that leak into the soil and groundwater, or that create toxic ash when incinerated. Using less packaging, and recycling more of it, is one of the great challenges for Green Consumers. (See "The Problems of Packaging," page 39, for more on this.)

❑ **Disposable Products.** In the post-World War II years, disposability became a status symbol, a measure of affluence for a mushrooming American economy. In 1955, *Life* magazine ran an article promoting "Throwaway Living." Its headline crowed: "DISPOSABLE ITEMS CUT DOWN HOUSEHOLD CHORES." But Americans' disposability has become more curse than blessing, representing the wastefulness that has contributed to much of our environmental mess. In the 1990s, disposability has become a paradox for many consumers: they want to be environmentally responsible, but they have gotten used to—some would say *addicted* to—the no-hassle, use-it-and-lose-it lifestyle. Example: We toss away 1.1 million tons of disposable plates and cups a year, enough to serve a picnic to everyone in the world six times a year! Manufacturers have responded to Americans' convenience cravings with an ever-impressive array of disposable technology: cups, containers, cameras, clothing, even computers. In fact, most of the products we buy are considered disposable when its useful life is over; never mind that the item may still be "useful" to someone else. Cheap parts and rising labor costs have made it easier to discard appliances such as toasters and TVs than to fix them.

Seeing the Big Picture

As you look at all these problems and the ways we contribute to them, you begin to see how many of the problems have multiple causes. Automobiles and electricity production, for example, are linked to several major environmental problems. Some of the same pollutants that contribute to global warming and the greenhouse effect also contribute to acid rain. Some of the ozone polluting products also clog landfills. Everything, it seems, is linked to something else. Trees, frogs, insects, clouds, microbes in the soil, plankton in the sea, humans—all are linked into one intricate and mutually supportive system. Disrupting one part of the system can cause devastation in some other part. Keeping the system healthy and intact is what the environmental—and Green Consumer—movements are all about.

THE THREE R'S

To help us remember what to do when shopping, the Green Consumer movement has its own set of "Three R's": refuse, reuse, and recycle. Each of them plays a key role in our attempts to minimize the environmental problems caused by our purchases and lifestyles. Keep in mind that this is a hierarchy: they are listed in descending order of preference.

1. Refuse

This is where the power of green is at its strongest. By refusing to buy wasteful and polluting products, you can make a powerful statement. As we have learned in other cases—most notably in the tuna industry's decision to go "dolphin-safe," and in McDonald's decision to switch from polystyrene to paper—a few Green Consumers can send shock waves through an industry.

What should you refuse to buy? Several things:

❑ products packaged in many layers of packaging
❑ products packaged in unrecycled or unrecyclable materials
❑ single-use and other products that have a short life before they must be thrown away

❑ products that are not energy efficient
❑ products made by companies with poor environmental records
❑ products purchased from retailers that have poor environmental records
❑ products that make false or misleading claims about their "greenness"

As we said before, few products are perfect. Your level of refusal to purchase some of these products will likely be influenced by the available alternatives. If a product you feel you need is available only in one form, and it is an environmentally undesirable one, you may choose to buy it anyway. But you don't have to accept this as "the best you can get." Consider writing the manufacturer and ask them to consider changing the product's packaging or contents to make it more environmentally responsible. Your letter will have more impact, however, if you have chosen not to purchase the product, and tell the manufacturer that. (See the "Action Guide" beginning on page 169 for information on writing to product manufacturers.)

"Refuse" is also known as "precycling," that is, eliminating wasteful products *before* you purchase them, rather than having to find a responsible means for disposing of them later on.

2. Reuse

Things that may only be used once before being thrown away are an inefficient use of our precious resources. It would be ideal if the products you do buy had the longest life possible. So, it is important to buy products that can be reused over and over. Consider batteries, for example. Why purchase batteries that must be thrown away—filling landfills with a melange of hazardous chemicals—when you can buy ones that are rechargeable hundreds of times? Why buy something that will have a short life when you can buy something that will last and last?

Some reusable things may cost a bit more to buy—rechargeable batteries, for example, are considerably more expensive than disposable ones—but over time, most of these products can more than pay for themselves. For example, a battery charger and four "AA" batteries sell for about $15, compared to four disposable "AA" batteries, which retail for around $3. But if you recharge the batteries only five times, you'll save enough money to recover the cost of the equipment, plus the electricity used for recharging. After that, you're ahead of the game!

Another aspect of "reuse" is to look for products made from or packaged in recycled material. By doing this you are supporting the reuse of resources. The greater the content of recycled material, the better. Some products or packages state specifically: "Made of 100% recycled content." Lacking such statements, it's difficult to tell the exact amount of recycled content.

3. Recycle

If you have refused and reused as much as possible, a high percentage of your leftover trash should be recyclable. And it is extremely important that you make sure to recycle what's left.

Recycling is not a new idea. During World War II, it was a way of life for Americans. Everything from tin cans to scrap iron, rubber to cooking grease, was recycled to help the war effort. Everyone did his or her part to preserve the country's scarce or strategic resources. But when the war ended and the economy boomed, there was a backlash: Americans were taught by advertisers that true prosperity meant having the luxury to use things once and throw them away. Unfortunately, that notion became a way of life for many people.

Now, recycling has come full cycle. Our scarce and strategic resources are once again being associated with some of our everyday purchases—the petroleum contained in the plastic packaging we buy, for example, the bauxite and energy used to produce aluminum, the drinkable water spoiled by the effluents of paper mills, and on and on. And so, recycling has once again taken on a new importance.

Consider just a few of the benefits recycling can bring:

❑ Recycling a single aluminum can saves enough energy to produce 20 more cans.
❑ Recycling a glass bottle or jar saves 25 percent of the energy it took to produce it and cuts up to half of the pollution created in manufacturing a new one.
❑ Using recycled paper instead of virgin paper for one print run

The Fourth "R": Reduce

How much is enough?

For many of us, the question may seem unanswerable. Since the end of World War II, we have been on a buying binge, acquiring and consuming goods at a remarkable rate. According to the Worldwatch Institute, American consumption since 1950 has soared. Per capita, energy use has climbed 60 percent, car travel more than doubled, plastics use multiplied 20-fold, and air travel jumped 25-fold.

And it's not just how much we consume, it's also the things we buy. From Buicks to beef to bottled water, many of our most popular purchases use a disproportionate amount of energy and resources. For example, while only 8 percent of humans, about 400 million, own cars, they are responsible for 13 percent of the world's emissions of carbon dioxide. While about one-fourth of the world's population eats meat, they consume almost half of the world's grain, which is used to fatten livestock. The meat eaters are also responsible for a greater portion of soil erosion, water pollution, and global warming resulting from raising livestock.

Is this right? A growing body of knowledge and thought suggests that a change may be in order, a move to simplicity, an attempt to consume less. Even some manufacturers and retailers are recognizing that consuming may not really be the road to happiness.

In 1990, for example, Esprit de Corps, the clothing retailer, launched a unique ad campaign urging consumers to ask themselves "before you buy something, whether this is something you really need." The company went on to explain, "It could be you'll buy more or less from us, but we'll be happy to adjust our business up or down accordingly, because we'll feel we are contributing to a healthier attitude about consumption."

Consuming less need not necessarily make our lives less comfortable. After all, how much packaging, junk mail, toys, and gadgets do we need? How much convenience is necessary? How much credit and debt can we afford?

It should be pointed out that despite the phenomenal growth in income and consumption, public opinion polls conducted since the 1950s consistently show that Americans are no more satisfied with our lives now than we were then. As much as we have, we still want more. Are we really better off than we once were?

of the Sunday edition of the *New York Times* would save the equivalent of 75,000 trees and reduce landfill waste.

❑ Each year, the amount of steel recycled in this country saves enough energy to meet the equivalent of the electrical power needs of the city of Los Angeles for more than eight years.

Despite these impressive statistics, Americans recycle only about a tenth of our household trash; another 10 percent or so is incinerated. The rest of the trash is tossed into landfills. In discarding our trash, we are also discarding vast amounts of raw materials—and the energy it takes to convert these materials into finished goods.

Is It Really Recyclable? Like just about everything else in this confusing green world, there are different levels of recyclability. Some materials are continually recyclable. For example, you can turn a recycled aluminum can into another aluminum can, and turn *that* can into still another can, and on and on. So too with glass bottles.

Other things have more limited recyclability. Paper, for example, can be continually recycled, but not necessarily into the same high grade as it was originally. So, a high-grade office paper may be recycled into a standard photocopy paper, perhaps even into a paper towel.

Still other things can be recycled only once, or a very few number of times. Most plastics fall into this category. Under current technology and sanitation laws, polystyrene (Styrofoam), plastic soda bottles, and milk jugs can be turned into other nonfood items—filling for ski jackets, videocassette casings, plastic "lumber," to name just a few possibilities. Some of the limits are imposed by federal laws that prohibit food to come in contact with most recycled materials. Those laws are being reviewed and are changing somewhat—it is now possible to make a foam egg carton containing up to 25 percent recycled material—but it is not likely that plastic will be transformed into a fully recyclable material in the near future.

Don't fall for manufacturer claims that a product or container is "recyclable." A growing number of companies have made this claim on products or labels based solely on the fact that the technology exists to recycle it, regardless of whether the technology is accessible to most consumers. So, a "recyclable" package for which there is no place in your community to recycle it is no better than an "unrecyclable" package. (See "A Short Course in Eco-Speak," page 31, for more on label claims.)

What Can You Recycle?

Not everything we buy can be recycled, but most things can. A study conducted on Long Island by the Center for the Biology of Natural Systems found that 84 percent of household waste—including food, yard waste, paper, bottles, and cans—can be recycled. Of course, the exact percentage of recyclable materials you throw away is related in part to what materials can be readily recycled in your community.

What exactly is recyclable? Here, in alphabetical order, is a status report on the recyclability of our most common products and materials.

❑ **Aluminum.** Aluminum is one of the most expensive and polluting metals to produce. For one thing, the metal is extracted from bauxite ore, which is mined on the surface, much of it in tropical rain forest areas. Fortunately, recycling aluminum cans is one of the great environmental success stories of recent years. Since the early 1970s, can recycling has grown steadily; Americans now recycle about 55 percent of all aluminum cans—some 42 billion recycled cans containing about 1.5 billion pounds of aluminum. According to the Aluminum Association, that represents more than 5 million tons of aluminum that has not been buried in our landfills since 1972.

The aluminum recycling process is relatively swift: Used aluminum cans are melted down and returned to store shelves in the form of new beverage containers in as few as six weeks. One reason for such speed is the ease of recycling an all-aluminum can: Because most aluminum cans have no labels, caps, or tops, they needn't be separated from other "foreign" materials before being reused.

Nearly all recycling facilities accept all-aluminum cans. (To tell the difference between an aluminum can and a steel one, apply a magnet. If the magnet sticks, the can is steel.) Most centers accept bimetal cans—those with aluminum ends and steel sides. In addition to cans, you may include aluminum foil, pie pans, and other aluminum scrap. You needn't prepare aluminum for recycling except to separate it from other trash. For your convenience, however, it might help to crush cans before discarding. Of course, in many areas, you can return the cans to stores for a refund.

❑ **Corrugated Cardboard.** Corrugated cardboard—used primarily in cardboard boxes—like paper, is fully recyclable. Cardboard, in particular, is in short supply, so the demand for used cardboard is high. About half of all corrugated cardboard used in the United States is recovered each year, and the percentage is expected to rise in coming years.

To recycle cardboard, you need only flatten boxes and tie them together for drop-off or pickup.

❑ **Glass.** Like aluminum, glass is 100 percent recyclable. Glass makes up about 10 percent of household garbage and is one of the easiest of materials to recycle. Broken glass (known as "cullet") is added to new molten glass in a furnace, producing new glass. Through the use of cullet, which melts at a lower temperature than the raw materials used to make glass, new glass can be made with considerably less energy and resources than would otherwise be needed. Glass can be reused an infinite number of times. Of course, some bottles needn't be melted down at all. Like milk bottles of years gone by, they are merely washed, sterilized, and refilled.

There are three basic types of glass: clear, green, and brown. Not all recycling centers accept all three types, but many do. In any case, you should separate the different color glass into different containers (or at least separate out the clear glass, if that's all your local recycling center will accept). Broken glass is perfectly acceptable, and while you needn't remove paper labels, you may need to remove aluminum neck rings and caps; your recycling center will provide details. Some types of glass aren't recyclable, however. Most recycling centers, for example, will not accept light bulbs, ceramic glass, dishes, or plate glass, because these items consist of different materials than do bottles and jars.

❑ **Household Batteries.** The number of dry-cell batteries we use is growing by nearly 10 percent a year. We now throw away about 2.5 billion batteries—from the cylindrical flashlight variety to the tiny button-size type used in cameras, watches, calculators, and hearing aids.

These innocent-looking batteries are rarely considered environmental threats, but they contain some very toxic chemi-

A Short Course in Eco-Speak

The Green Consumer movement has produced its share of misleading and confusing terminology, much of which has been overused and abused by advertisers and marketers. All of the terms below are vague and have no legal definitions; in fact, few experts agree on what they mean. You should be wary of these phrases. Here are a few to watch out for:

❑ **"Degradable," "Biodegradable," "Photodegradable"**: Technically, *everything* degrades eventually, even if it takes thousands of years. But in a landfill, which is where most of our trash ends up, few things degrade. Landfills tend to preserve trash better than they dispose of it. These terms aren't any more meaningful when referring to cleaning products. There is little agreement among scientists about how quickly or how thoroughly an ingredient must break down in water before it can be called "biodegradable."

❑ **"Green"**: This means nothing by itself. As we've pointed out, it is a relative term that can have many, many meanings. When used as a marketing or labeling term, it has no more meaning than any other color of the rainbow.

❑ **"Natural"**: This is a widely overused and abused phrase that has little meaning. There are many "natural" ingredients — lead, for example — that are extremely poisonous.

❑ **"Nontoxic"**: Again, there is no legal definition of this term. Substances that are not poisonous to people may be poisonous to plants, animals, insects, or bacteria in the soil.

❑ **"Ozone-friendly"** or **"Won't Harm the Ozone Layer"**: This usually indicates that a product is made without ozone-destroying chlorofluorocarbons. But that doesn't make a product environmentally benign. Foam coffee cups, for example, may no longer be made with CFCs, but they are still wasteful. Most aerosol products don't use CFCs either, but have other environmentally harmful ingredients.

❑ **"Recyclable"**: Lots of things can be recycled, but not everyone can recycle them. Something is "recyclable" only if you can — and will — recycle it in your community. If you don't have the ability to recycle something (or if you simply don't bother) a "recyclable" package or product is no better than an "unrecyclable" one. See also "Recycled" below.

❑ **"Recycled"**: Some "recycled" paper contains only 5 or 10 percent recycled content, while other products have 100 percent recycled content. There is also a difference between "post-consumer" waste (thrown away by consumers after use) and "pre-consumer" or "post-industrial" waste (products manufactured but never sold, or scraps swept off the factory floor, for example). So, "recycled" itself needs additional information to be meaningful. See also "Recyclable" above.

❑ **"Safe for the Environment"** or **"Environmentally Safe"**: *Nothing* is safe for the environment. Everything has some environmental impact. This phrase is simply untrue.

cals. When burned in incinerators, the heavy metals in batteries—including cadmium, lead, lithium, manganese dioxide, mercury, nickel, silver, and zinc—pollute the air or become toxic components of discarded incinerator ash. When tossed into landfills, the metals leach out of corroded batteries and seep into the groundwater. These metals are so dangerous that the Occupational Safety and Health Administration has established workplace exposure limits for all eight metals mentioned above.

Battery recycling is done routinely in Europe and Japan, but it has only begun to catch on in the United States, which has only a handful of battery recycling programs.

Another solution comes in a new breed of mercury-free disposable batteries. While they're still thrown away, at least they lack one of the most toxic ingredients. Eveready Battery Company has introduced an alkaline battery that is 99.975 percent mercury-free. Another source is Power Plus of America (1605 Lakes Pkwy., Lawrenceville, GA 30243; 404-339-1672).

Perhaps the best solution are rechargeable batteries. Although these batteries cost more and require the one-time added expense of a recharger, most such batteries can be recharged up to one thousand times. After that, the nickel and cadmium can still be reclaimed and recycled. There are even solar-powered battery charging systems, which use sunlight to recharge batteries. It takes about six hours to recharge a set of batteries. Seventh Generation (800-456-1177) sells one such system.

❑ **Motor Oil.** While an increasing number of people are changing their own motor oil, many of them—as well as a number of professional service stations—do not dispose of the used oil properly. And if you think that the few quarts of oil draining from your car's crankcase won't hurt anyone, think again: As little as a quart of oil, when completely dissolved or dispersed in water, can contaminate up to 2 million gallons of drinking water. A single gallon can form an oil slick of nearly eight acres. Used motor oil poured down the drain or into the nearest storm sewer often goes directly to the nearest creek, river, or lake, killing aquatic life and polluting drinking water.

Many communities have begun curbside collection programs to recycle used oil. Used oil is a valuable, renewable resource, although it must be handled carefully. Through refin-

Paper or Plastic?

You may be getting weary of the debate over paper versus plastic—grocery bags, that is—but the last skirmish has yet to be fought. In fact, the claims are getting more aggressive—and more confusing.

The issue is which type of bag—paper or plastic—is less polluting and, therefore, the checkout choice for green consumers. Plastic has generally been avoided by Green Consumers, despite manufacturer claims of being bio- or photodegradable. Those claims, of course, have been found untrue; very little, it turns out, degrades in a landfill. And paper bags have been undesirable because of the significant amounts of air and water pollution created by the paper industry. (Because paper bags were made of weak recycled fibers, they simply weren't strong enough to hold a bagful of groceries.)

But today, supermarkets are setting up plastic-bag recycling programs, and paper bag recycling technology has improved. Suddenly, things have gotten more complex.

New Bags From Old News

One big change has been the addition of recycled paper bags that can withstand the rigors of shopping. Several companies are now making bags containing up to 100 percent recycled fibers, some from old newsprint. One company, Duro Bag, makes a 100 percent recycled paper bag. Duro Bag's product is made of 70 percent post-consumer waste and 30 percent post-industrial waste.

On the plastic side, a growing number of recycling programs have cropped up, with collection bins placed in supermarkets. But plastic bags can't be made into more plastic bags easily because many recycled bags contain food wastes or leftover receipts, which hamper the recycling process. Moreover, plastic-bag recycling does not appear to be a break-even proposition, meaning its future in the marketplace is uncertain.

What about using no bag at all? Cloth tote bags have become big business; many supermarkets sell them. But cloth bags don't appear to be taking America by storm, most likely because we tend to make large shopping trips, unlike Europeans, for example, who make small purchases daily. Still, totes are the best, albeit least practical, solution.

Does it really matter? There are those who believe that the great bag debate is irrelevant. "I think there is a danger when people ask, 'Should I use paper or plastic?' because there is an assumption that somehow the material matters," says Jeanne Wirka, solid waste source reduction specialist at Environmental Action Foundation. "It's not the material, it's the fact that it's a disposable product that matters. What you really have to do is change your habits. The only basis on which you should choose one or the other is if you are going to recycle them."

ing, about two and a half quarts of new motor oil can be extracted from one gallon of used oil. When recycled, used oil is reprocessed with water, then used for fuel. About a fourth is re-refined and turned into base oil stock. With additives, it is used as lubricating oil and put to other uses. Some of it is used for road oil, dust control, wood preservatives, and fireplace "fire log" ingredients. The production of re-refined oil uses just one-fourth the energy of refining from crude oil. According to the U.S. Department of Energy, Americans use about 1.2 billion gallons of oil annually—about 78,000 barrels a day. About 60 percent of it (some 700 million gallons) consists of used motor oils. The remaining 500 million gallons are industrial oils.

Many neighborhood service stations and auto repair shops will accept your recycled oil. (They may have strict standards on how it should be packaged and delivered to them so be sure to ask before bringing in your oil.) A growing number of service stations and auto parts retailers also accept used motor oil, including many **Jiffy Lube** shops, all **Pep Boys** stores, many **Chief Auto Parts** stores, and all **Sears** auto centers.

❑ **Paper.** The United States leads the world in paper consumption—and trails far behind in paper recycling. Americans use some 67 million tons of paper annually, recycling about a fourth of it. This statistic is particularly distressing in light of the fact that the production of a ton of paper from discarded waste paper requires 64 percent less energy, needs 58 percent less water, results in 74 percent less air pollution and 35 percent less water pollution, saves 17 pulp trees, reduces solid waste going into landfills, and creates five times more jobs compared to producing a ton of paper from virgin wood pulp. According to the Institute of Scrap Recycling Industries, more than 200 million trees are saved each year due to current recycling efforts. Paper makes up nearly a third of municipal solid waste by weight and well over half by volume. Each ton of paper not recycled uses three cubic feet of landfill space and incurs as much as $100 in disposal costs.

The lack of paper recycling in America has less to do with people's unwillingness to recycle than with the lack of recycling mills in operation. In 1990, only nine of the 43 newsprint mills in the United States were capable of using recycling fibers; most

Which Plastic Is It?

How can you tell the kind of plastic used in a package? It can be very hard to do. In fact, most experts we talked with said that it's nearly impossible to tell without conducting a a chemical analysis.

Some help comes from a coding system created by the plastics industry. This voluntary system was created to help recyclers identify which packages could be recycled. If you can recycle plastic in your comunity, this system will help you identify packages that are recyclable. (If a container does not have a code, you can safely assume that it is made of several types of plastic, and is therefore not recyclable.) Please understand: *This system does not ensure that your plastic purchase will be recycled.* It is simply an identification system.

The system includes seven codes, one of which is usually stamped on the bottom of a package inside a recycling logo like the one on page 37. Please note that of these seven plastics, only two of them—PET and HDPE—are currently being recycled in any quantities. Here are the seven codes:

1. Polyethylene terephthalate (PET)—used mostly in soft-drink bottles, but also found in meat containers, cosmetic packages, and boil-in bags.

2. High-density polyethylene (HDPE)—the most common plastic in consumer products, including milk, water jugs, shampoo, and detergent bottles.

3. Polyvinyl chloride (PVC or V)—used to package floor polishes, shampoos, edible oils, mouthwashes, and liquor.

4. Low-density polyethylene (LDPE)—used when squeezability is desired, such as in toiletries and cosmetics.

5. Polypropylene (PP)—used to package foods that must be packaged while hot, such as pancake syrup.

6. Polystyrene (PS)—used in egg cartons and to package tablets, salves, ointments, and other products not sensitive to oxygen and moisture.

7. Other plastics—includes a wide range of substances, including mixed plastics that cannot be recycled.

mills run at full capacity and sell all the recycled paper they can produce. But the demand for recycled paper continues to outstrip supply and all indications are that the supply-demand imbalance will remain. Building a new plant can cost $450 million and can take several years; even converting an old plant to accept recycled paper can cost $80 million.

Not all paper is recyclable. Newsprint is among the most desirable papers, but recyclers are picky even about that. Coated papers—such as those used for Sunday magazines and newspaper advertising supplements—can make sheets of rolled paper stick together. Yellow paper, including legal pads, is also undesirable. Brown bags and junk mail also gum up the works. The widespread assumption that brown paper is always recyclable simply isn't true. It must first be separated from other types of paper.

A few state-of-the art plants, however, can recycle many of these types of paper, as well as envelopes with plastic windows, greeting cards on glossy stock, and other things. The Fort Howard Corporation in Wisconsin, for example, uses old magazines, phone books, Post-its, and many other types of supposedly unrecyclable paper to produce its two lines of paper towels, toilet paper, and napkins.

Recycling paper is easy. Newspapers may be kept in a separate bin, then tied (with cotton string) into bundles a foot or two thick for easy handling. Nearly all recycling plants accept bundled newspapers.

Most office papers may be recycled too. Office workers produce an average of a pound of waste paper a day, according to the Office Paper Recycling Service, a New York consulting group. (Ironically, much of the paper is generated by computers and other equipment that was supposed to represent the paperless office of the future.) Unfortunately, the common yellow legal pad is not accepted by most paper recyclers because the dyes make it difficult to produce the bright white paper most consumers want; other paper colors are equally undesirable.

With the demand for recycled office paper being much greater than the demand for newsprint—and, therefore, the price for recycled office paper being considerably higher than for newsprint—many large companies have recognized that recycling office paper is as profitable as it is environmentally

How to Tell If It's Recycled

Finding recycled products at the market takes a bit of investigative work. Very few products boldly announce that their products are made from recycled material, and there have been a couple of instances in which companies' claims have been misleading.

Recycled packaging can be identified in three ways:

❑ The recycling symbol on the package (pictured here).
❑ A statement such as "This package made from recycled materials."
❑ A gray interior in paperboard boxes, such as those used for cereals, detergents, and cake mixes. A white interior usually indicates that the package is made from virgin materials. (However, a box with a white interior may still be made from recycled material. To check, tear a corner of the package; if you see gray, it is made from recycled paper.)

sound. According to published reports, the American Telephone & Telegraph Company saved $1 million in disposal costs in 1988 and made a $365,000 profit by recycling its office waste paper. Other companies report similarly impressive figures. The World Trade Center in New York City set up a program for its 50,000 workers. It is expected to save 262,500 gallons of water, 153,750 kilowatt-hours of electricity, 371 cubic feet of landfill space, and 637 trees—*every day*.

❑ **Plastics.** As stated earlier, plastics recycling is a growing—and controversial—process. Researchers have developed processes to shred plastics, primarily from soda bottles, into flakes, which are then washed and separated from residual materials and made into a wide range of materials. But unlike paper, glass, and aluminum, at present, plastics can be recycled a limited number of times. Moreover, only a relatively small number of recycling centers and even fewer curbside pickup programs currently accept plastics. If you can recycle plastic, it is best to rinse it, then crush it to save valuable space in the vehicles that haul the plastic to be recycled.

❑ **Steel Cans.** All steel food and beverage cans are 100 percent recyclable. Yet, each year, about 30 billion steel cans with a thin tin coating are dumped into America's landfills. The technology to reclaim and recycle these two materials has been around for more than sixty years, and the capacity to recycle far exceeds the availability of recyclable cans. You can easily determine which cans contain steel with a simple magnet: if the magnet sticks, it's steel; if not, it's probably aluminum.

❑ **Tires.** We throw away 220 million tires a year, millions of which end up in huge, unsightly—and fire prone—stockpiles around the country; currently, well over a billion discarded tires sit in such mounds. But tires can be recycled in a number of ways: almost 40 million tires are retreaded annually; another 10 million are shredded, then used for sheet rubber, asphalt-rubber for roadbeds, roofing material, and other products; some shredded tires are used as fuel to generate electricity. You can purchase flooring tiles made from recycled tires. Some people recycle tires themselves by using them in gardening to protect tomato and other fragile plants.

❑ **Yard Wastes.** Leaves, cut grass, and other yard wastes represent about 20 percent of all waste that ends up in landfills—about 35 billion tons a year. Yet these materials can be of use, and their recycling can save money. Yard wastes are usually recycled into compost, an organic substance that can be used to make soil richer for growing things. Composting takes place when yard wastes and other matter decompose under controlled conditions. Anyone can compost. You can buy simple composting devices from nurseries, gardener's supply companies, and from mail-order companies such as Seventh Generation. (See "Shopping by Mail," page 182, for names and addresses.)

Zap!

Still another packaging concern is what happens to some packages when used in a microwave oven. There is evidence that chemicals contained in some of the most common types of microwave packaging may "migrate" into foods when exposed to the high temperatures of microwaves. The Food and Drug Administration has conducted tests which found that certain chemicals consistently leaked into food during cooking. One FDA tests found that some frozen food containers intended for cooking in *conventional* ovens also leaked chemicals into food. Even the standard "cling wrap" many consumers use to cover leftovers before reheating in a microwave are suspected of migrating chemicals into foods during cooking. (Fatty foods are a particular problem: A British study found that cling wraps can leak chemicals into fatty foods at room temperature and even during refrigeration.) The bottom line is that there are no legal definitions for such terms as "microwave-safe," "microwave-approved," and even "microwaveable." Their use does not mean that the containers can be safely used in a microwave oven.

The best bet when microwave cooking is to cover the dish with glass. Wax paper works just fine to cover a dish. If you must use a plastic container or wrap, don't let it touch the food.

The Problems of Packaging

By now, you've probably heard some of the amazing eco-statistics about packaging: that we go through 2.5 million plastic bottles *every hour*, or that we discard enough glass bottles and jars to fill the 1,350-foot twin towers of New York's World Trade Center *every two weeks*. Each statistic seems more incredible than the next. But they are based on reality.

To most consumers, packaging seems an attractive, protective wrapper at best, a crass and wasteful indulgence at worst. But to packaging experts, it is a combination of art and science. How do you get a perishable product to market and into consumers' homes while keeping the product sanitary and in tip-top shape? How do you protect against the new breed of terrorists known as tamperers—those who alter the contents of foods, toys, or medicines with harmful or even deadly ingredients? How do you provide sufficient information about the product so as to induce shoppers to choose it in favor of the competition?

And increasingly: How do you accomplish all this in a way that has the least impact on the earth and its resources?

The fact is, a great deal of packaging is unnecessary and wasteful. You needn't walk very far down any supermarket aisle to find something like this: a plastic bowl covered with a plastic lid, contained in a cardboard box, which is shrink-wrapped in still more plastic. Ironically, some of these over-packaged goods are given awards by the packaging industry for their innovative designs. It is precisely these "innovations" that contribute to our clogged landfills and other environmental problems. Of the roughly two tons of trash discarded by the average American each year, packaging accounts for an estimated 30 percent, or about 1,200 pounds for every man, woman, and child. By some estimates, packaging accounts for almost two-thirds of what we throw away every day.

The New Generation of Packaging. The problem isn't just the amount of packaging, it's also the types of materials being used. A growing number of products are being wrapped in "composites"—packages containing several layers of materials and adhesives, such as aseptic juice boxes, which contain layers of polyethylene, paperboard, and aluminum. Squeezable ketchup and mustard, made of up to seven layers of plastics and adhesives, are another example. When you are finished with such packages, you cannot separate the various materials for recycling, so they are guaranteed to end up in landfills, where they will last for centuries, or incinerators, where their ash may contain toxic materials that can get into the air, soil, and water.

Even when packaging consists of only one type of material, it is often an unrecyclable one. The vast majority of Americans have no means to recycle most types of plastic or polystyrene, or even the kind of coated paperboard used on cereal and cracker boxes and many other packages. Many manufacturers, trying to lure environmentally conscious consumers, claim their packages to be "recyclable." As stated earlier, while technically true, it is not true for most consumers.

The environmental problems of packaging aren't limited to what happens to it when you throw it away. The manufacturing process is also key. The amount of air and water pollution created during the package's manufacture, the amount of energy required to transport it to market, the need to keep it refrigerated or frozen (requiring more ozone-depleting coolant)—all are factors in determining which material or combination of materials comprise the best types of packaging.

And then there are the inks used in printing the packages' colorful labels. Some of the inks contain heavy metals—toxic elements such as lead, cadmium, and chromium. Although studies are inconclusive, some researchers believe that heavy metals may leach into groundwater when packages are discarded in landfills, or that they may contaminate the ash that spews from incinerators when the packages are burned. Heavy metals are just the beginning. Dyes, solders, and other additives are under scrutiny for their possible environmental impact.

Definitive information about all these issues is elusive. While each industry—glass, paper, aluminum, plastic, polystyrene, and juice boxes—boasts impressive studies supporting the notion that theirs is the material of environmental choice, there is little consensus, even among independent scientists.

The Best and Worst Packaging

Here are the basic packaging categories, listed from best to worst. In reviewing this list, keep in mind that a key component is the type of material that can be readily recycled in your community. If you live in a community in which paperboard cereal boxes, for example, or plastic soda bottles can be recycled, your list of "best" and "worst" will differ from someone who cannot recycle these things.

1. Best. What's the best type of packaging? Easy: no packaging at all. You'd be surprised how many products don't need any packaging. Is it really necessary to shrink-wrap produce on a cardboard or Styrofoam tray? Probably not. (In fact, you probably don't even need plastic produce bags for fruits and vegetables with their own peels or rinds. You throw five cans of soup into your grocery cart and shopping bag, why not five oranges?) Many nonfood items also can be sold perfectly well without packaging. Where sanitation and security are not a concern—as with hardware and variety store items—packaging may not be needed.

2. Very Good. When some packaging is required, the rule is: Use the least amount of packaging possible, using the highest content of recycled and recyclable material. So, a glass bottle with an aluminum or steel top—all of which are easily recyclable—is perfectly adequate. An aluminum, steel, or bimetal can is fine, too. (A bimetal can combines aluminum, tin, and steel; they are used to package many foods.) Seeking minimal packaging also leads one to buy the largest-size package possible. Several smaller packages inevitably mean a greater amount of packaging materials for the same amount of product. Another way to minimize packaging is to buy one of the growing number of concentrated products, especially detergents and other cleaners; simply put, they combine more product in less space. Some brands combine detergent and bleach, thereby halving the number of packages needed to do laundry. Both are good strategies for cutting packaging.

What about cardboard? Many products, including most cold breakfast cereals, are packaged in materials made from recycled paper or cardboard collected from large, industrial users of these materials. That's the good part. But many recyclers do not accept these for recycling, so they will end up in incinerators or landfills.

Also ranking high are endlessly refillable packages—namely returnable glass beverage bottles. Egg cartons, assuming they are returned and reused, also rate high.

3. Good. Reusable, refillable, and recyclable packages are a step in the right direction. Margarine tubs and other containers that can be used for food storage are good examples, although there are limits to the number of these most households can put to use. Another innovation are products in which you purchase the big package once, then buy smaller refills from there on. Concentrated Downy Refill, for example, which comes in a nonrecycled and nonrecyclable milk carton-like container, can refill a hard plastic jug of the original product; you simply pour in the concentrate and add water, like making frozen orange juice. Pillsbury sells a microwave cake mix that includes a reusable plastic pan. After buying the pan, you need only buy refills from there on. Eventually, of course, the cake pan—and the empty fabric softener jug—will need to be discarded. Nei-

Best to Worst

Best	❑ No packaging at all
Very Good	❑ Minimal, recycled, and recyclable packages
	❑ Concentrated products
	❑ Endlessly refillable packages
	❑ Reusable, refillable, and recyclable packages
Good	❑ Packages made of recyclable, but not recycled, material
	❑ Packages made of recycled, but unrecyclable, material
Fair	❑ Packages made from partial recycled content
	❑ Products packaged in multiple layers of recyclable packaging
Bad	❑ Products packaged in multiple layers of unrecyclable packaging
	❑ Single-serving containers
	❑ Aseptic packages
	❑ Aerosol cans
Worst	❑ Packages made of composite materials

ther type of container is easily recyclable.

Of course, anything that you can recycle is desirable. But as stated previously, those materials vary from community to community, even neighborhood to neighborhood. Better still are recyclable packages made from recycled material.

4. Fair. In the middle ground are packages made from recyclable, but not recycled, material. Again, this will have a lot to do with what's recyclable in your community. Less desirable are packages made from recycled materials that cannot themselves be recycled; they are at the end of their useful life and destined for a landfill or incinerator. Still okay but barely acceptable are containers made from at least some recycled material. Some plastic jugs, for example, contain up to 25 percent recycled plastic. You can find polystyrene egg cartons also containing one-fourth recycled polystyrene.

5. Bad. Products packaged in several layers of wrapping are usually not desirable, and their desirability drops in direct proportion to the amount of unrecycled and unrecyclable material used. Supermarket shelves are being increasingly filled with such products, mostly convenience and microwave foods. But not exclusively: we found overpackaged frozen foods, snacks, beverages, cleaners, baby products, and personal care products.

Equally undesirable is single-layer packaging made from unrecycled and unrecyclable materials. Again, specific products in this category have a lot to do with what you can recycle in your community. For some people, plastic soda bottles and milk jugs will fall into this undesirable category because they cannot recycle these things. For those who can recycle these plastics, such beverage containers would probably fall into the "Fair" category above.

Some of these "Bad" products look innocent enough. A white cardboard box surrounding a box of donuts, for example, may not appear villainous, but that packaging may take up as much space in a landfill as a polystyrene meat tray, may be just as polluting to manufacture, and may take nearly as many centuries to degrade.

6. Worst. At the bottom of the heap—literally—are single-serving containers, especially when they are made from multiple layers of materials, also known as "composites." Single servings, no matter what they're made of, create a great deal of packaging waste for the amount of product delivered.

It's not that Green Consumers must abandon convenience in their daily lives. But there are greener alternatives: resealable plastic containers can hold small servings of just about anything, including beverages. Moreover, if necessary, you can place these containers in a microwave oven for cooking. (**Beware**: *Remove the tight-fitting plastic top and cover with wax paper before cooking; otherwise, heat will not be able to escape and the container will melt.*) An investment in a set of such containers will more than pay for themselves through the cost savings you'll get from buying large-size packages, instead of single servings.

Multiple Materials. Composites are particularly troublesome. Squeezable ketchup and mustard containers, for example, made from several layers of plastics and adhesives, are simply not recyclable, even in communities that have the ability to recycle several kinds of plastic. They are guaranteed to end up in landfills.

Another environmentally undesirable container is the aseptic package, also known as the juice box—those small boxes of juice and flavored drinks, accompanied by their own plastic straw. Juice boxes are made from three layers of materials—70 percent coated paperboard, 23 percent polyethylene plastic, and 7 percent aluminum foil. While they're lightweight and less voluminous, taking up minimal space in a landfill, they are not readily recyclable. (The aseptic industry has managed to turn some juice boxes into a kind of plastic lumber, but this is a demonstration project only and currently has no economic feasibility for wide-scale use.) Again, there are alternatives, including small resealable plastic juice containers that can be washed and reused. By buying larger sizes with less packaging, you'll get far more beverage per dollar—as well as doing something good for the environment.

Aerosols. Spray cans are another undesirable packaging type. Many people believe that because chlorofluorocarbons were banned in 1978, aerosol cans are now environmentally safe products. But aerosols aren't environmentally safe. For starters, many of the substitute propellants—which force the hair spray or deodorant out of the can—are hydrocarbons, such as propane or butane. Just like the hydrocarbons that come out of your car's tailpipe, these mix with sunlight to form photochemical smog. For that reason, Los Angeles will soon be regulating the sale of things like aerosol hair spray and deodorants. In addition, some common household aerosols contain toxic solvents that end up in the air and in our lungs. And aerosol packaging itself is unrecyclable; because you can't separate the various components of the can, they are virtually guaranteed to end up in landfills.

Making the "right" packaging choices can be extremely difficult. If you have kids, for example, it may be difficult to give up juice boxes. But do what you can. By starting somewhere, you will be making a small but meaningful contribution to reducing our overflowing landfills and helping to make the most of our resources.

Pesticides and Poisons

"For the first time in the history of the world, every human being is now subjected to contact with dangerous chemicals, from the moment of conception until death. In the less than two decades of their use, the synthetic pesticides have been so thoroughly distributed through the animate and inanimate world that they occur virtually everywhere. They have been recovered from most of the major river systems and even from streams of ground water flowing unseen through the earth. Residues of these chemicals linger in soil to which they may have been applied a dozen years before. They have entered and lodged in the bodies of fish, birds, reptiles, and domestic and wild animals so universally that scientists carrying on animal experiments find it almost impossible to locate subjects free from such contamination. They have been found in fish in remote mountain lakes, in earthworms burrowing in soil, in the eggs of birds—and in man himself. For these chemicals are now stored in the bodies of the vast majority of human beings, regardless of age. They occur in the mother's milk, and probably in the tissues of the unborn child."

Rachel Carson wrote these words—in 1962, in her landmark book *Silent Spring*. Three decades later, worldwide pesticide sales have more than quadrupled to more than four billion pounds a year. In the United States, American farmers' use of herbicides, insecticides, fungicides, and other toxic chemicals have increased tenfold since the end of World War II. And many of these chemicals are far more toxic than those written about by Carson.

Consider a few statistics:

❑ American farmers use about 1.5 billion pounds of pesticides each year—five pounds for every man, woman, and child. These pesticides may end up in as many as half the foods we eat.
❑ Yet only about 1 percent of all food shipments are tested for pesticides. In fact, standard government laboratory tests can't even detect more than half of the 500 or so pesticides currently in agricultural use.
❑ According to the Environmental Protection Agency, 66 pesticides sprayed on food crops contain cancer-causing agents. Of

the 560 million pounds of herbicides and fungicides used by U.S. farmers each year, the EPA says that about two thirds probably cause cancer.

❑ According to the National Academy of Sciences, only 10 percent of the 35,000 pesticides introduced since 1945 have been tested for their effects on people.

❑ There are about 50,000 pesticide-related illnesses in the United States each year.

❑ According to a report issued by the U.S. Public Interest Research Group, common foods are treated with dozens of possible carcinogens: 25 in corn, 24 in apples, 23 in tomatoes, and 21 in peaches, for example.

❑ About 40 percent of pesticides are used primarily to make food look good, according to a study by the California Public Interest Research Group.

For years, many people thought that all these chemicals were simply the price we must pay for an abundant food supply. But a growing body of knowledge has led many experts to question the need for the chemical-intensive farming that has become the norm in American agriculture.

Some pesticides, such as Alar, commonly used on red apples, have received widespread publicity as carcinogens. But Alar is only the beginning. As the U.S. Public Interest Research Group revealed, apple crops are regularly dusted with a cornucopia of chemicals: benomyl, captafol, captan, chlordimeform, dicofol, dinoseb, folpet, lead arsenate, mancozeb, maneb, methomyl, metiram, o-phenylphenol, parathion, permethrin, phosmet, pronamide, silvez, simazine, tetrachlorvinphos, thiophanate-methyl, toxaphene, and zineb.

Poisoning People. Those at particular risk from pesticides are children. According to a comprehensive two-year study released in 1989 by the Natural Resources Defense Council, pesticide tolerance levels established by the federal government are set only for healthy adults. Children face special danger: because they weigh less, they eat proportionately more pesticide-containing foods than adults. They also eat more fruit, which makes up an estimated one-third of preschoolers' diets, compared to one-fifth of adults'.

Farmworkers also are threatened by these chemicals. Workers routinely suffer nausea, vomiting, stomach cramps, headaches, and lowered levels of important blood enzymes. A National Cancer Institute study found that workers exposed to herbicides are at six times greater risk of contracting cancer of the lymphatic system. Where fields are crop dusted—sprayed with chemicals by low-flying airplanes—sprayed chemicals often drift into fields miles away, affecting workers. According to the *Journal of Economic Entomology*, an average of 54 percent of aerially applied insecticides drift off their intended target. The remainder is carried into areas for which spraying was not intended. (Even when sprayed from the ground under ideal conditions, 3 to 5 percent drift into the air.)

It would be enough if pesticides affected only people. But they also wreak havoc on many other parts of the environment. For example, crop-dusting's drifting chemicals threaten waterfowl and other aquatic life. Agricultural run-off—irrigation and rain water that runs off farm land and into surface and ground water—contaminates the drinking water supply for about half of all Americans.

Fertilizers are another problem. One study found that nitrogen-based fertilizers increase the levels of methane in soil. Methane is one of the more potent gases contributing to global warming.

Back to the Future. The movement to reduce pesticide use in farming has caught fire. What was once the domain of a handful of "alternative lifestylers" has hit America's bread basket. Adopting methods known as "sustainable agriculture," thousands of farmers concerned about the environment—and the rising cost of chemicals—are trying to reduce their chemical habits. As the movement grows, some experts predict as much as a 30 percent reduction in chemical use on American farms.

Sustainable agriculture doesn't eliminate chemicals altogether. Instead, it combines limited chemical use with time-tested natural techniques and some high-tech ones: crop rotation, the introduction of natural predators that feed on unwanted pests, genetic engineering of pest-resistant plants, and sanitary practices to reduce breeding areas for pests. But the chemical reductions are impressive. For example, cultivating an

Thought for Food

Here are some things to keep in mind about organic produce:

- ❑ Organic produce sometimes isn't as attractive as nonorganic produce. But its imperfections do not negatively affect taste or quality. The nature of organic farming is that while produce isn't always uniform and blemish-free, it can be tastier and better for you.
- ❑ Organically grown food tends to be locally grown, using less energy and creating less pollution, compared with produce shipped from long distances.
- ❑ Organic certification does not mean that a product is entirely free of chemical residues. Air pollution, drifting sprays from nearby farms, and persistent soil toxins can all contribute small amounts of pollutants.

acre of bulb onions takes nearly 40 pounds per acre of fungicides, insecticides, herbicides, and other chemicals. Using integrated pest management techniques requires less than half that amount and may eliminate some ingredients altogether, depending on weather and other conditions.

Farming Without Chemicals

Some farmers are going cold turkey, eliminating chemicals altogether to grow organic crops. Organic farming isn't new, of course, but it has changed radically over the years, both in growing techniques and in its acceptance as an agricultural technique.

Organic farmers don't try to conquer nature so much as work with it, preventing the need for "intervention" with the use of unwanted chemicals. Today's organic farmers work by a specific set of standards that can be verified and certified by on-farm inspections, soil tests, and annual management reviews. There are about 30 private and public certification organizations in effect. The 1990 Farm Bill established the mechanism to create national certification standards for organic food.

Is organic food healthier than nonorganic food? There is much debate on this. Skeptics note that an apple is an apple, no matter what coaxed it out of a tree. The nutritional value of the organic apple (or any other produce) should be virtually identical to the nonorganic one. Still, some nutritionists claim that organically grown food has superior nutritional value.

Nutrition aside, it is the lack of toxins—and the lower impact on the environment—that gives organically grown food an advantage over traditional food. With the continuing sagas of contaminants in food—Alar in apples, aldicarb in watermelons, and EDB in grain, among others—organically grown food, if not more nutritious, is less risky, both to you and the earth.

How to Tell If It's Organic. How can you make sure that something truly is organic? The best sources of information are the certification labels carried on product packages and crates. There are four principal organizations, all of whose mission statements are similar:

❑ **California Certified Organic Farmers** (Box 8136, Santa Cruz, CA 95061; 408-423-2263) calls itself "an organization of farmers and supporting members working to promote and verify organic farming practices, and supporting all efforts for a healthier, sustainable agricultural system."

❑ **Farm Verified Organic** (P.O. Box 45, Redding, CT 06875; 203-544-9896) describes itself as "an internationally accepted farm-to-table product-guarantee program that determines the authenticity of organically grown food."

❑ **Organic Crop Improvement Association** (Box 729A, New Holland, PA 17557; 717-354-4936) is a self-described "internationally recognized, brand-neutral, farmer-owned crop improvement, quality assurance program, which is backed by an independent third-party certification and audit-controlled system."

❑ **Organic Growers and Buyers Association** (1405 Silver Lake Rd., New Brighton, MN 55112; 612-636-7933) is a longtime organic certifier, which prides itself on "commitment to quality standards."

Each group has its own seal, which can appear only on products meeting each organization's standards. Those standards tend to be strict. All produce must be raised using acceptable products and techniques, and violators are subject to having their entire *field* disqualified for a number of years. Such requirements point up the longevity that many chemicals have in the soil; the life of these chemicals does not end when the crop is harvested.

As usual, things aren't black and white (or green and ungreen). Besides organic, there are also "transitional" produce and grains, which have been grown without artificial chemicals, but on land that has not yet been totally cleansed of previous chemical use.

The **Organic Foods Production Association of North America** (OFPANA, P.O. Box 31, Belchertown, MA 01007; 413-774-7511) is a trade organization for the organic farming industry. Members include growers, processors, wholesalers, distributors, and retailers. OFPANA will supply its "Organic Farmers Associations Council," a list of organic associations in the U.S.; send a self-addressed, stamped envelope. It may help you find where organically grown food is sold in your area.

Another source of information on where to find organically grown food is **Americans for Safe Food** (ASF, 1501 16th St. NW, Washington, DC 20036; 202-332-9110), a project of the Center for Science in the Public Interest. ASF publishes a list of reliable mail-order companies that will ship their organic products directly to individual consumers, usually via United Parcel Service; most do not require minimum orders.

To obtain the most recent list, write to ASF at the address above. Include $1 and a stamped, self-addressed legal-size envelope with 55 cents postage.

Finding Organic Food in Supermarkets. Organic food has been available mainly through health food stores and so-called "farmers' markets." With the rise in organic food production has come the tentative acceptance of organics by mainstream supermarket chains.

But only tentative. Consumer surveys indicated an increased demand for organics, leading stores to stock them. But consumers haven't necessarily backed up their demand with dollars. Organic produce has languished in many stores, causing some chains to drop it like a hot potato. Organic growers' organizations have complained that supermarkets didn't give their products the necessary care or promotion, while supermarkets have accused growers of not being able to provide a steady supply of salable product.

But organically grown produce may not *seem* salable to many Americans because it is not always as uniform in size, shape, and appearance. Factory farming methods, whatever their effect on the environment, have managed to produce perfect looking fruits and vegetables all year long. Organic methods, because they don't use some of the cosmetic-improving chemicals of traditional growing methods, often are less attractive, although they would likely taste the same or better to a blindfolded sampler. So appearance may have much to do with organics' lack of success in the mainstream marketplace.

Eco-Labeling

The ultimate solution for all of our Green Consumer dilemmas would be to have one commonly accepted seal of approval on packages, certifying that a product had met a set of standards for environmental responsibility. With such a label, we could more easily decipher advertising claims, choose recycled and recyclable packaging, and avoid products containing toxins and other pollutants.

But the world just isn't that simple—least of all the world of Green Consumerism.

It isn't that no one is trying. Indeed, there are several competing attempts to devise a credible "eco-label," or at least a set of generally accepted definitions and standards. Such labels have been in place in other countries—including Canada and Germany—with mixed success. In the United States, two private organizations have created product-labeling programs, the Coalition of Northeast Governors has tried to establish some standards, as has a task force of state attorneys general. And the federal government—through Congress and the Federal Trade Commission—keeps threatening to get into the act.

All indications are that things will become more complicated before they become simpler.

Green Cross and Green Seal

S.C.S.
Green Cross
CERTIFIED
TM
PRODUCT
These RENEW bags are certified to be made from at least 80% recycled plastic.

The two most visible programs are the two private ones:

❑ **Green Cross Certification Company** (1611 Telegraph Ave., Ste. 1111, Oakland, CA 94612; 415-829-1415) was created by an independent testing company to review products and award seals to those that meet the state of the art. Areas of interest include packaging, biodegradability, energy efficiency, and evaluating the claims of so-called "sustainable resource" products.

❑ **Green Seal** (1733 Connecticut Ave. NW, Washington, DC 20009; 202-328-8095) is the creation of Denis Hayes, who also was a guiding light for Earth Days 1970 and 1990. It plans to look at the entire "cradle-to-grave" life cycle aspect of products; the product categories will be similar to those on Green Cross's list.

Will they work? Even in their initial stages, there was much contention by each organization about the other's approach: Green Cross has claimed that conducting credible life cycle analyses is not yet possible; Green Seal has criticized Green Cross for what it claims is a piecemeal approach to eco-labeling. But both programs have good leaders with good credentials—and high ambitions.

Stay tuned.

What Makes a Company Green?

Products aren't the only things that can be considered "green." Companies—those that make products and those that do other things—also can be green or ungreen, or any of several shades in between.

The fact is, companies can have an enormous impact on the environment. And not just heavy industries, the ones notorious for belching pollutants from smokestacks, or dumping filthy waste water into rivers and oceans. Every company has an impact on the environment, from big conglomerates to Mom-and-Pop operations, including:

❑ the products they make and the services they perform
❑ the way they operate and maintain their facilities and equipment
❑ the relationships they have with suppliers and customers that can encourage or discourage wasteful behavior
❑ the relationships they have with their employees, and even their families, which can encourage or discourage them to be environmentally responsible citizens
❑ the relationships they have with the communities in which they operate that can encourage or discourage communities to adopt policies that are good for the environment

One key aspect of being a Green Consumer is to patronize companies with good environmental records and forward-thinking policies, and to boycott companies that have products and policies that are bad for the environment.

Granted, it is not always easy determining a company's "greenness." Many companies are so big that it is difficult to determine the many things they do. And many products—particularly those sold in supermarkets—are sold under brand names that are several steps removed from the parent company. For example: The Jolly Green Giant is owned by Pillsbury Co., which itself is a subsidiary of Grand Metropolitan USA. The fact that companies and brands seem to be continually bought, sold, merged, and acquired does not make it any easier.

There are few perfectly green companies, just as there are few perfectly green consumers or products. Nearly every com-

The Greenest Companies

Below are some of the most environmentally responsible companies, based on data compiled by the Council on Economic Priorities. Note that CEP's data do not track many of the smaller companies that have fine environmental policies and practices, so its data base contains a higher percentage of larger firms.

• **Abbott Laboratories**—its chemical and agricultural divisions have recycled, reused, or otherwise minimized 90 percent of its 800,000 gallons of hazardous waste. It donated nearly 1,000 hours of service and $55,000 to a wetland restoration project in Illinois.
• **Aveda Corporation**—makes cruelty-free products, and has started an in-salon recycling program for returning empties.
• **Church & Dwight**—developed an "environmentally sound" paint-stripping compound that removes paint without using toxic chemicals.
• **Clorox**—has developed an emission-control and release-prevention program and has been a leader in using recycled packaging. It has taken responsibility for cleaning up its on-site toxic wastes.
• **H. J. Heinz**—set a "zero-tolerance" for killing dolphins when acquiring tuna for its StarKist brand and has spent millions on researching recyclable plastic bottles. It also has pledged not to use irradiated foods, ingredients, or spices.
• **Kellogg**—has implemented waste management and recycling programs around the country; at the Battle Creek plant, it reduced landfill scrap by 60 percent. It completed a five-year program to replace solvent-based printing inks to meet EPA clean-air standards.
• **Melitta North America**—has produced unbleached coffee filters and those using an alternative bleaching process; over 90 percent of its packaging is recycled. In 1990, the company donated two trees for every one it used in filter production.
• **Tom's of Maine**—makes the only nationally available toothpaste packaged in a 100-percent aluminum tube in a recycled paperboard box; it has refused to produce an unrecyclable pump dispenser.

pany has some ungreen skeletons in its closet. Moreover, there are companies in which one division makes a good, green product, while another division has a bad environmental record. Example: Ultracel, an innovative furniture seat cushion made without using chlorofluorocarbons, was introduced in 1989 by Union Carbide, the company responsible for the Bhophal disaster of 1985, in which some 2,000 people died from a chemical leak. Another example: Procter & Gamble, whose disposable

How to Be a Greener Company

Like people, each company is unique. And just as each individual has different ways of "going green," companies, too, have a variety of options available to them. What's right for one company may not work for another.

Just as we tell people not to try to be perfectly green — lest they get frustrated and discouraged at the enormity of it all — we tell the same things to companies. It's hard to do everything right, even if you are making an environmentally benign product or performing a service that doesn't seem to pollute a thing. Just being in business is a polluting activity. There is so much paper, so many purchases, so much energy and other resources needed just to operate from day to day.

Here are a few things companies can do to be more environmentally responsible. For further reference, consult *50 Simple Things Your Business Can Do to Save the Earth*, by the Earthworks Group, available for $8.45 postpaid from Tilden Press, 1526 Connecticut Ave. NW, Washington, DC 20036.

❑ **Conduct a waste audit.** A lot of the things companies throw away are recyclable. Paper, of course, which makes up about 41 percent of landfills, is the first thing to look at. Office paper use has grown enormously in recent years, despite repeated promises of a paperless "office of the future." But it has become painfully obvious, as one expert put it, that the paperless office is about as practical as a paperless bathroom. But paper is only the beginning. Cardboard boxes, packaging materials, aluminum, glass, steel, and some plastics all can be saved from the waste stream.

There is good incentive for companies to do these things. Many companies are earning a tidy sum selling the "trash" they used to throw away. (In fact, these companies

diaper products, Pampers and Luvs, are responsible for about 1 percent of all household solid waste in the U.S., is also one of the biggest innovators of green products and packages.

Green Guidelines. Still, there are some basic questions anyone can ask that can help judge a company's concern for the environment:

❑ Does the company recycle and encourage others to do so?
❑ Is the company in compliance with federal, state, and local environmental laws and regulations?
❑ Does the company use or support use of alternative energy

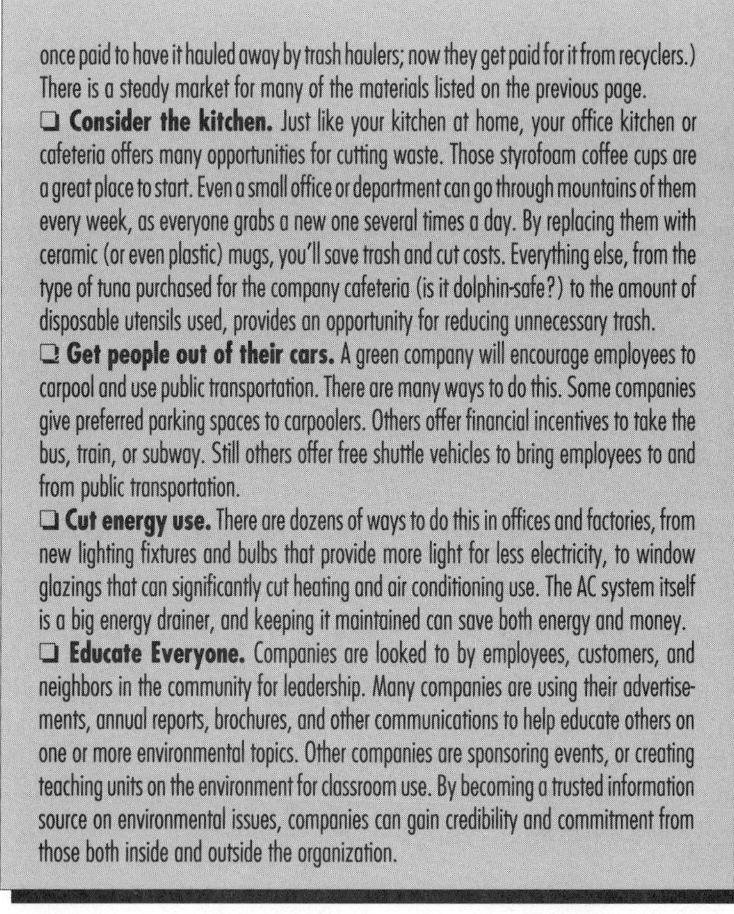
once paid to have it hauled away by trash haulers; now they get paid for it from recyclers.) There is a steady market for many of the materials listed on the previous page.

❑ **Consider the kitchen.** Just like your kitchen at home, your office kitchen or cafeteria offers many opportunities for cutting waste. Those styrofoam coffee cups are a great place to start. Even a small office or department can go through mountains of them every week, as everyone grabs a new one several times a day. By replacing them with ceramic (or even plastic) mugs, you'll save trash and cut costs. Everything else, from the type of tuna purchased for the company cafeteria (is it dolphin-safe?) to the amount of disposable utensils used, provides an opportunity for reducing unnecessary trash.

❑ **Get people out of their cars.** A green company will encourage employees to carpool and use public transportation. There are many ways to do this. Some companies give preferred parking spaces to carpoolers. Others offer financial incentives to take the bus, train, or subway. Still others offer free shuttle vehicles to bring employees to and from public transportation.

❑ **Cut energy use.** There are dozens of ways to do this in offices and factories, from new lighting fixtures and bulbs that provide more light for less electricity, to window glazings that can significantly cut heating and air conditioning use. The AC system itself is a big energy drainer, and keeping it maintained can save both energy and money.

❑ **Educate Everyone.** Companies are looked to by employees, customers, and neighbors in the community for leadership. Many companies are using their advertisements, annual reports, brochures, and other communications to help educate others on one or more environmental topics. Other companies are sponsoring events, or creating teaching units on the environment for classroom use. By becoming a trusted information source on environmental issues, companies can gain credibility and commitment from those both inside and outside the organization.

sources, waste reduction techniques, and nonpolluting technologies?

❑ Has the company made an effort to reduce unnecessary materials use and to buy recycled products whenever feasible?

❑ Does the company dispose of waste in an environmentally sound way?

Granted, these are not the only good questions to ask, but their answers will provide a reasonable indicator of the company's interest in and commitment to the environment.

While it is nearly impossible for most of us to track the environmental records of companies whose products we buy, there is a growing effort to follow companies' records on the environment. One of the most ambitious efforts is being conducted by the Council on Economic Priorities (CEP), a New York-based nonprofit group. For more than 20 years, CEP has been keeping track of companies' records on the environment and other things. Their findings are published each year in a handy guidebook, *Shopping for a Better World.* The book rates thousands of products based on companies' records on philanthropy, military contracts, community outreach, and affirmative action, among other factors. CEP's company environmental ratings are the same ones that appear in this book.

In compiling its ratings, CEP rated large companies differently than small ones. It gave high marks to large companies with "positive programs, such as the use and encouragement of recycling, alternative energy sources, waste reduction, etc. A record relatively clear of major regulatory violations." For small companies, CEP looked for those that "Make strong efforts to 1) use biodegradable and/or recyclable materials in packaging, 2) dispose of waste from manufacturing processes in an environmentally sound way, and 3) use only natural ingredients and growing techniques for food." Large companies rated at the bottom were those that have "a poor public record of significant violations, major accidents, and/or a history of lobbying against sound environmental practices." Small companies that rated poorly were those that made "little or no effort to implement proactive programs" or that had "significant regulatory infractions."

Another CEP project is the Environmental Data Clearinghouse, established in 1990 to gather information on corporate environmental performance and make that information available to a wide range of individuals and organizations.

For more information on EDC contact the Council on Economic Priorities, 30 Irving Pl., New York, NY 10003; 212-420-1133. Annual membership ($25) in CEP includes a copy of *Shopping for a Better World* and monthly research reports. *Shopping for a Better World* is also available in bookstores, or from CEP for $6.45 postpaid.

Part Two

A·L·O·N·G T·H·E S·H·E·L·V·E·S

About the Ratings

These days, rating products using environmental standards involves equal parts of both art and science. Part of the problem has to do with the fact that the necessary science doesn't yet exist. Even the experts don't truly understand the cradle-to-grave impact of many materials and manufacturing processes. That information is only beginning to emerge.

What's more, there are tens of thousands of supermarket products out there, not all of which are available in every store. Some of these products come in a variety of sizes, shapes, and packaging materials. Sometimes, a product sold in one region of the country will appear with a different package, a different set of ingredients, or even a different name in another region. On top of all this, companies add, drop, or reformulate their products at the drop of a hat. There are hundreds of such changes every week.

While such a rich and changing mix of products makes America a great place to live and shop, it also makes it a difficult marketplace in which to create a standardized set of environmental ratings.

In the pages that follow, we have not attempted to rate every product on the market. Besides being nearly impossible to do, that list would be quickly outdated and of minimal value. In some cases, there would be few if any differences among the products. Among cold breakfast cereals, for example, there are few differences in packaging and contents. The same is true of many other categories of packaged goods.

Instead, we have attempted to give an environmental tour of the supermarket aisles, offering information and insight into how to view your purchases through a "green" lens. In addition, we have provided names of hundreds of products we think you should either seek out or avoid, based on the products' packaging, contents, and on the environmental records of their manufacturers. In Part Four of this book, we have provided an Action Guide, with addresses and phone numbers for these and other product manufacturers, so that you can register your comments about their products or contact them for more information.

As we've said before, few products are perfectly "green," and that includes most of the products to which we've given favorable ratings in this book. Rather than seeking unattainable perfection, we considered the state of the art, comparing each product to the highest known standard. If, for example, the best way currently to package a powdered laundry detergent is in concentrated form in a recycled and recyclable paperboard box, we used that as the standard for detergent packaging. In a few cases—the contents and packaging of air fresheners, for example—the state of the art was sufficiently poor, environmentally speaking, that even the "best" was not worth recommending.

We must point out that we are not scientists, nor have we conducted laboratory analyses of these products. Instead, we have relied upon examinations of packaging types, research into ingredients, interviews with experts, label information, product data sheets, data bases about companies' environmental records, and other journalistic and research techniques. The data on companies were provided by the Council on Economic Priorities (CEP), a New York-based nonprofit research organization, which also publishes *Shopping for a Better World*, a best-selling guide to corporations' records on a variety of social issues. (See page 58 for more information about CEP's ratings, and about how to obtain a copy of its book.)

With each listing of best or worst products, we've included a symbol that indicates the category or categories that warranted the rating:

Best Choices

means a product rated highly for its contents

means a product rated highly for its packaging

means a product's manufacturer rated highly

Worst Choices

means a product rated poorly for its contents

means a product rated poorly for its packaging

means a product's manufacturer rated poorly

You should be aware that some of the more environmentally responsible products currently on the market may not be found on any supermarket shelves. Rather, they are available through health food stores and through any of the growing number of mail-order companies specializing in green products. (See "Shopping by Mail," page 182, for information about these companies.) We have included many of these products in our ratings.

But that does not mean that supermarkets *shouldn't* carry these items. Indeed, they will—if you ask them to. Store owners and managers tell us repeatedly that they try to be responsive to customer requests. If even a handful of shoppers ask for a certain type of product—say, for paper towels made from recycled material, or for phosphate-free cleaning products—they will make every effort to stock them. (Of course, it's important that you back up your requests with action: If few people actually buy these products once they become available, the store will stop carrying them.) So, if your store doesn't stock a green product you are looking for—ask.

A final word. While we have tried to list as many specific products as possible, we must emphasize that what's most important is to learn the rules of the game. By understanding the various environmental impacts of the many different things you buy, you will be well armed to be a Green Consumer—no matter where you shop, or how often companies change their products.

Beverages

We drink a lot of beverages. In fact, we drink more soda than water—about 47 gallons of soda a year for every man, woman, and child, compared with about 37 gallons of plain tap water. Add juices, coffee, tea, and all of the exotic blends that seem to be cropping up on shelves each week and suddenly the beverage section has become a major shopping experience by itself.

What's the Problem?

Along with all these new products have also come new types of packaging. It's no longer just bottles and cans. There are pouches and boxes and bags, some of which are made of several layers of materials that cannot easily be recycled.

We've already talked about the best and worst types of packaging. When it comes to beverages, you can find examples at both ends of the spectrum. At one end is the aluminum can or glass bottle—a readily recyclable container in a "closed-loop" system, meaning that the can be turned into another can, and the bottle into another bottle.

At the other end, there are such products as Wylers' Big Squeeze Fruit Drink, consisting of six eight-ounce bottles made of several layers of different plastics, with snap-off plastic tops and paper labels in a cardboard container—all of which is shrink-wrapped in yet more plastic! A similarly packaged product, Squeezit, was named 1989's Beverage Packaging of the Year by *Food and Drug Packaging*, a trade magazine—demonstrating just how out of touch the food industry often is with Green Consumers.

Juice Box Jive

Another award-winning beverage container is the aseptic package, also known as the juice box. You can find them everywhere—those little rectangular boxes with their built-in straws. They're lightweight, convenient, and nearly indestructible.

Juice boxes are made of three layers of materials: paperboard, polyethylene plastic, and aluminum foil. It's this thin combination that gives juice boxes their special lightweight convenience. Unfortunately, these same properties make it extremely difficult to recycle juice boxes.

If all we intended to do with our trash was to send it to landfills, juice boxes would probably be a good idea, because they take up very little space when crushed. But as most Green Consumers know, landfilling is not the goal. Nearly every city, county, and state has announced plans to recycle a growing amount of our garbage, reducing the need for landfills and getting the most use out of all of our resources.

Juice boxes have been recycled in very small quantities, in programs subsidized by juice box manufacturers. They can be turned into a type of plastic "lumber," among other things. But such recycling will not be available to many people for years, if at all. So despite industry's claim, juice boxes are, for all intents and purposes, not recyclable.

Fortunately, there are substitutes for juice boxes. Several juice and drink products now come in small aluminum or steel cans, or glass bottles, all of which are readily recyclable. But buying lots of small cans or bottles isn't particularly green. The best bet is to invest in a Thermos or a set of small reusable plastic beverage containers. They can be sent to school or work—and brought home again and refilled, over and over.

What to Look For

❑ Choose products made from 100 percent recyclable material. In most cases, that's glass and aluminum, although in some communities it may also include PET or HDPE plastic bottles.

❑ Look for products that don't have extra layers or plastic "yokes" (which hold together six-packs of bottles or cans) or plastic layers of shrink wrapping. If you do buy packages with yokes, be sure to cut them up into small pieces before discarding; marine animals have been known to get caught in the yokes' loops.

❑ Buy the largest-size container of a product as possible, avoiding juice boxes and other smaller containers. If you or your kids prefer small portable containers, buy a set of plastic bottles that can be used over and over. (If you can't find them in your supermarket or kitchen supply store, check with a store that sells camping equipment. Backpackers and other campers swear by plastic containers with tight, leakproof lids.)

❑ Stay away from powdered beverages that come in individual packets. In most cases, such products—cocoa mix, for example—are also available in larger jars or cans. (True, you'll have to mix your own cocoa, but how do the manufacturers know exactly how chocolatey you like your cocoa anyway?)

❑ If your community recycles PET plastic, stay clear of colored plastic bottles unless you are sure they will be accepted by your recycler. (Of course, if you can't recycle PET plastic, stick with aluminum or glass containers.)

❑ Look for products from local bottling companies. They require the least amount of energy and pollution to get from manufacturer to market.

❑ Choose products from companies that have good environmental records.

The Bottom Line

As stated earlier, we found a wide range of beverage products and packages on the shelves. Many of them were packaged responsibly, and several brands—namely, Deer Park Spring Water, Hershey, H.J. Heinz, Melitta, Eden Foods, Stokely-Van Camp, R.W. Knudsen, Santa Cruz Natural, and Celestial Seasonings—received high marks for corporate environmental responsibility.

The key here is simplicity: Look for the product with the greatest amount of beverage or mix, and the least amount of everything else.

Bottled Water

Once the beverage of a small but dedicated corps of purists, bottled water is now the beverage of choice for millions of Americans. We drink nearly 2 billion gallons a year, at an average cost of about $1.15 per gallon (compared with an average price of $1.28 per *thousand* gallons of tap water—that makes bottled water nearly 900 times more expensive).

Part of the popularity of bottled water has to do with the

image of purity—images of water drawn from carefully protected sources in high mountain regions of the United States and Europe. But in many cases, such images are largely the product of clever advertisers. The fact is, some bottled water isn't much purer than tap water.

Even the most popular bottled water products go through a fairly complex manufacturing process before being shipped. Some waters get their carbonation from gases shipped in from another state. Some products themselves are shipped by truck or train hundreds of miles from source to bottler. With such factory-like conditions, it's not surprising that occasionally, a handful of bottles will become contaminated with chemicals (as Perrier was in 1989, when a few bottles were found to contain benzene, a dangerous chemical).

If you are drinking bottled water for its purity, remember this: There are few state or federal laws that require bottled water to be any purer than the stuff that comes out of your kitchen or bathroom faucet.

What about the environmental impact of bottled water? For one thing, the resources used in packaging (usually in plastic, it seems) and the energy and pollution resulting from shipping these products around the country and overseas, are tremendous, given that most of us have adequate resources on hand. So, bottled water wastes energy resources and contributes to air pollution, global warming, acid rain, and other environmental menaces.

Given all this, bottled water is not a particularly green product, no matter how many pollutants you feel you may be avoiding by drinking it. And because it represents a waste of packaging and transportation resources, we can't truly recommend any brand of bottled water.

If you believe your tap water is a problem, consider buying water from nearby sources, packaged in recyclable containers. Your best bet may be to rent a dispenser that holds refillable five-gallon glass or plastic jugs of water. There are many delivery services that will bring refillable containers of water to your home on a regular basis; check the Yellow Pages for listings.

Other than buying domestic over imported bottled water, and the other recommendations on pages 65-66, no brands of bottled water we found rated as a "best" or "worst" choice.

Cocoa and Milk Modifiers

Watch out for hot cocoa mixes that place eight or ten pouches in a box. These pouches are usually unrecyclable, made from several layers of materials. Look instead for simple glass jars or cardboard boxes, the type of packaging we rated as "best."

Best Choices

Carnation's Malted Milk (Carnation Co.)
Droste Cocoa (Droste USA Ltd.)
Hershey's Chocolate Milk Mix (Hershey Foods Corp.)
Hershey's Cocoa (Hershey Foods Corp.)
Ovaltine (Sandoz Nutrition Corp.)

Worst Choices

Carnation Hot Cocoa Mix (Carnation Co.)
Carnation Sugar Free Hot Cocoa Mix (Carnation Co.)
Nestle's Cocoa (Nestle Foods Corp.)

Coffee

Coffee has grown from a simple product into a high-status and high-tech item. Not that many years ago, the choice was simply between ground and instant. Ground coffee, more often than not, was made in a percolator; instant coffee was, and still is, a matter of adding boiling water to a teaspoonful or two. How things have changed!

What's the Problem?

Today, making a cup of coffee is partly a matter of choosing the right "technology"—drip, automatic drip, plunger, or good old-fashioned percolator, among others. Concern over caffeine consumption has led to a rise in decaffeinated coffees, some of which use chemicals similar to those used in dry cleaning to remove caffeine from coffee beans. There is no information available on the environmental impact of such chemicals. (There are alternative processes available to decaffeinate coffee beans, notably water-processing.)

Packaging. The way coffee is packaged is another matter. While ground coffee traditionally comes in a recyclable steel can, several brands, emulating the way coffee is sold in gourmet shops, package grounds in vacuum-sealed, foil-like pouches. These pouches, however, are composites, made of several layers of materials, and are not recyclable. (On the other hand, some supermarkets sell coffee grounds or beans in paper pouches, an environmentally desirable package.)

Even worse are all-in-one filter packs, individual packages intended to eliminate measuring and pouring grounds into a coffee filter. These are needless and wasteful.

Deforestation. Some coffee comes from countries that are losing their forests at a rapid rate, although the links between coffee production and deforestation aren't yet clear enough to recommend cutting down on brands from particular countries. The problems are most likely to occur in Brazil, Central America, and West Africa.

What to Look For

- ❑ Buy ground coffee in a steel can or paper bag.
- ❑ Buy instant coffee in a steel can or glass jar.
- ❑ Avoid premeasured coffee packs.
- ❑ If you buy decaffeinated coffee, look for the water-processed variety.

See also information on coffee filters under "Paper Products."

Best Choices

Any brand packed in a steel can or glass jar.

Any Melitta brand (Melitta North America Inc.)

Worst Choices

Any brand packed in a vacuum-sealed plastic package.

Maxwell House Filter Packs (Kraft General Foods Inc.)

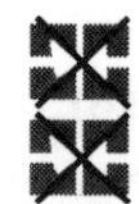

Health and Diet Drinks

There are difficult choices here. Soy milk—a nondairy beverage popular among those who cannot drink cow's milk—is usually packaged in aseptic packaging (juice boxes). Yet one of the leading soy milk companies, Eden Foods Inc., is highly rated for its environmental record. The result is a trade-off between badpackaging and a good company. The best choices below are largely those packaged in recyclable glass containers.

Best Choices

Eden Organic Barley Malt (Eden Foods Inc.)

Kaffree Roma (Worthington Foods Inc.)

Necta Soy Non-Dairy Beverages (Natural Inc.)

Sego Liquid Diet Meal (Sego)

Slender Diet Meal for Weight Control (Carnation Co.)

Worst Choices

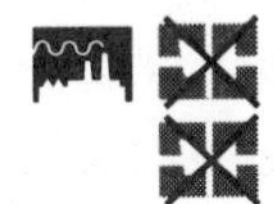

Edensoy (Eden Foods Inc.)

Sweet 'n' Low Instant Tea (Cumberland Packing Corp.)

Juices

Unfortunately, aseptic packages seem to have taken over this aisle; even worse, many juice boxes are also wrapped in plastic. Still, there are many juice products packaged in glass bottles and steel or aluminum cans. Some brands come as both juice boxes and as bottles or cans. When buying juice, fresh and canned are better than frozen. The amount of energy it takes to freeze and ship frozen juice is substantial. Ounce for ounce, getting frozen orange juice to market takes four times the energy of providing fresh oranges, according to the Worldwatch Institute.

Best Choices

After the Fall (bottles) (After the Fall Products Inc.)

Gatorade (bottles and cans) (Stokely-Van Camp Inc.)

Hollywood Natural Carrot Juice (Hollywood Foods)

Libby's Nectars (Carnation Co.)

Ocean Spray Cranberry Juices (bottles) (Ocean Spray Cranberries)

R. W. Knudsen products (R. W. Knudsen & Sons)

Recharge Tropical Thirst Quencher (R. W. Knudsen & Sons)

Santa Cruz Natural Juices (Santa Cruz Natural)

Tropicana Fruit Juices (Tropicana Products Inc.)

Veryfine Juices (Veryfine Products)

Welch's Juices (glass bottles) (Welch Foods Inc.)

Worst Choices

After the Fall (aseptics) (After the Fall Products Inc.)

Capri Sun 100% Natural Punches (Caprisun Inc.)

Del Monte juices (Del Monte USA)

Dole juices (Dole Food Co.)

Gatorade (aseptic) (Stokely-Van Camp Inc.)

Hawaiian Punch (aseptics) (Del Monte USA)

Hi-C Punches (aseptics) (Coca-Cola Co.)

Juicy Juice (Nestle Foods Corp.)

Kool Aid (aseptics) (Kraft General Foods Inc.)

McCain Junior Juice (aseptics) (McCain Citrus Inc.)

Minute Maid 100% Natural Punches (aseptics) (Coca-Cola Co.)

Mott's Juices (aseptics) (Mott's USA)

Ocean Spray Cranberry Juice (aseptics) (Ocean Spray Cranberries)

Squeezit (General Mills)

Welch's Juices (aseptic) (Welch Foods Inc.)

Wyler's Big Squeeze Fruit Drink (Thomas J. Lipton Co.)

Powdered Mixes

There are those who like their beverages dry—at least when they buy them. Powdered drink mixes have become popular items, and in at least one way they are ecological: By eliminating water, they reduce the amount of weight that needs to be transported from manufacturer to market. Why ship a pot's worth of tea when a couple of tea bags will allow you to make the product yourself at home?

Unfortunately, many of these mixes are packaged poorly. In striving for convenience—companies apparently believe we prefer only premeasured and individually wrapped packages—manufacturers have created some packaging monsters. Consider Crystal Light, a product of Kraft General Foods: The mix comes packaged in a plastic tub with a foil cover. The tubs are inside a high-density polyethylene plastic container with a plastic lid and a plastic safety seal. There are four or six such tubs in every product. Need we say more?

There are simpler versions of many of these products, including simple, recyclable glass jars. That's what garnered the "best ratings" below.

Best Choices

Lipton Iced Tea Mixes (Thomas J. Lipton Co.)

Nestea (Nestle Food Corp.)

Worst Choices

Alpine Instant Spiced Cider Apple Flavor Drink Mix (Krusteaz)

Carnation Instant Breakfast (Carnation Co.)

Country Time Lemonades (Kraft General Foods)

Crystal Light (Kraft General Foods)

Wyler's Fruit Slush Freeze and Eat (Thomas J. Lipton Co.)

Sodas

We guzzle a lot of soda—about 47 gallons a year for every American. And we go through a lot of soda bottles—some 2.5 million plastic soda bottles *every hour*, according to one estimate. Indeed, soda bottles have become a symbol of our throwaway society. They litter the highways, they crowd our landfills.

The good news is that many of these same containers are being viewed as one of the great environmental success stories of the 20th century. The fact that in just two short decades we have come to recycle some 55 percent of aluminum beverage cans is remarkable; much of that growth has taken place in just the past three years. There have been similar, albeit more modest gains with glass bottles.

What's the Problem?

Plastic bottles, however, do not share in this good news. Only a very small percentage of these bottles, made of polyethylene terepthalate (PET—often signified by a numeral "1" inside a recycling logo on the bottle's bottom) are being recycled. That may change, as more and more communities collect these bottles, and more plastic recyclers accept bottles.

But even if plastic bottles were recycled at the same rate as glass and aluminum, they would still be at a disadvantage. While, as we've previously said, a glass or aluminum beverage container can be turned into another glass or aluminum beverage container, a plastic bottle can only be turned into another plastic commodity—fiber fill for a ski parka, for example, or fence posts or parking-space bumpers. All of these will eventually need to be disposed of. In late 1990, Pepsi and Coca-Cola both announced plans to introduce recycled PET bottles, although they had not yet received government approval to do so.

(Another problem with PET plastic recycling has to do with the current lack of recycling plants. The company that recycles about two thirds of America's PET plastic must ship the used bottles to plants in the East Coast and in Europe. Clearly, such long-distance transportation adds considerably to the energy and pollution associated with plastic packaging.)

What to Look For

❑ Look for containers that are readily recyclable in your community. In most cases, that will be glass or aluminum; in some communities, it will also include PET plastic bottles.

❑ Avoid extra packaging, such as plastic shrink-wrapping or yokes around six-packs. Look for bottles and cans packaged in recycled cardboard containers.

❑ Avoid products that have foam labels. A simple paper label is all you need.

The Bottom Line

Nearly every soda manufacturer packages its products in several formats: aluminum cans in cardboard boxes; cans in plastic yokes; glass bottles in boxes, plastic yokes, and plastic wrapping; bottles with paper or foam plastic labels; plastic bottles with hard-plastic bases; and so on. And no company was rated either high or low for environmental responsibility.

So, drink what you like, packaged as simply as possible, and make sure to recycle the empties.

Tea

Because hot tea has become a favored drink among natural food buyers, there are many organic products available. Some of these, like Celestial Seasonings' products, also use tea bags that have not been treated with chlorine bleach. (Chlorine is linked with trace amounts of dioxin, a cancer-causing poison, in bleached paper products. See "Paper Products" for more on this.)

But don't assume that "natural" and organic products are responsibly packaged. For all their "naturalness," most products come in unrecycled cardboard boxes, which are wrapped in plastic. Yogi Tea Co. is in the same league with Lipton: both use a nonrecycled, shrink-wrapped cardboard box, then wrap each individual tea bag in paper. We hope that Yogi will meditate on reducing some of that unnecessary trash.

Best Choices

Bigelow Teas (R. C. Bigelow Inc.)

Boston Teas (Boston Tea Co.)

Celestial Seasonings (Celestial Seasonings Inc.)

Worst Choice

Any individually wrapped tea bag.

Dairy Products

There's nothing more natural than a herd of dairy cows grazing on lush green grass. Unfortunately, the modern dairy industry is anything but natural—it's a modern milk factory. Close to 98 percent of all the milk we drink comes from these factory farms, where what are advertised as contented cows are in fact treated as four-legged milk machines.

What's the Problem?

Dairy technology has become complex and high-tech, aimed at getting maximum milk for minimum cost. Cows are given antibiotics, hormones, and tranquilizers. Residues of some of these ingredients can be found in the milk itself. According to the experts, the concentrations of these residues are so weak as to be harmless.

Milk production and distribution isn't completely harmless to the environment. The factory-farm uses vast amounts of water and energy. But because milk is produced locally, transportation is relatively minimal.

Packaging. The clear glass milk bottle has all but disappeared, as has home delivery of milk and dairy products. (Not completely, though; you can still find milkmen—and milkwomen—in many small towns.) Most of us buy milk in paperboard cartons or plastic jugs. While most people would assume that paper is better than plastic, this may be one case where plastic wins out. The reason: Paper milk cartons are not recyclable. They are destined for landfills and incinerators. Plastic jugs, made from high-density polyethylene (HDPE—#2 in the plastic industry coding system), are being recycled in limited numbers. It is also a limited form of recycling: Government regulations do not yet allow recycled jugs to be made into more milk jugs, or any other type of food container. However, some recycled HDPE is being used to make plastic jugs for laundry detergent and other nonfood products.

What about other dairy products? For the most part, yogurt, cottage cheese, sour cream, and other products are packaged in plastic, and there is extremely little deviation in packaging style

Best Choices

Any cheese in a single layer of packaging.

Any yogurt not packaged in plastic.

Stonyfield Farm brand yogurt (Stonyfield Farms Inc.)

Weight Watchers brand (H.J. Heinz Co.)

Worst Choices

Light 'n' Lively yogurts (Kraft General Foods Inc.)

Mini Bon Bel cheeses (Fromageries Bel Inc.)

Nabisco products (Nabisco Brands Inc.)

Polly-O String Cheese (Kraft General Foods Inc.)

Yoplait Light 4-pack (Yoplait USA Inc.)

from company to company. Cheese may be one exception. We found some brands with individually wrapped slices, a convenience for those who either lack the time to slice cheese themselves, or who perhaps can't cut straight.

What to Look For

❑ Look for minimally packaged products. Avoid those with individually wrapped slices and servings.
❑ Buy the largest quantity of a given product that you can to minimize packaging.
❑ Look for products packaged in cardboard instead of plastic. While all national brands we found used plastic, some local dairies still use cardboard packaging.
❑ Check to see if you can recycle HDPE plastic milk jugs in your community. If so, buy milk in those containers.

Frozen Foods

For many of us, answering the question "What's for dinner?" is as simple as opening the freezer door. But frozen entrees—what we used to call "TV dinners"—are far from simple when it comes to assessing their environmental impact. Not that these heavily packaged and processed products could rightfully be called "green." But accurately determining the environmental impact of frozen foods compared to other foods isn't easy to do.

The frozen-food business is hot, annually ringing up more than $20 billion worth of frozen meat, vegetables, desserts, and just about anything else that's edible. And they're not just for dinner anymore: frozen breakfasts are growing even faster, up over 20 percent a year.

What's the Problem?

Let's start with the obvious: Locally grown fresh food is hands-down more ecological than processed food of any kind, including frozen. It uses little, if any, packaging and the least amount of energy to get to the marketplace. Except for food scraps, there is little waste from fresh food. Still, fresh food transported cross country and from other countries can be more energy-intensive than frozen foods, according to one study.

Having said that, let's deal with the reality that fresh foods aren't always convenient or available to a lot of people. And some folks can't cook, are too busy to cook, or would at least enjoy a change of pace.

Energy and CFCs. Frozen food, as it turns out, fares reasonably well in energy consumption compared with canned or other packaged foods. A 1977 study of frozen versus canned peas, for example—still considered one of the better studies to date—concluded that the cradle-to-grave energy costs were equal. True, you needn't freeze canned peas all the way from warehouse to stove-top, but making tin cans turned out to be energy-intensive, too. However, the 1977 study was sponsored by the Frozen Foods Institute, an industry association. While the study has not been disputed, the jury is still out on this question.

Energy isn't the only factor. Most freezers—industrial models in warehouses, commercial types in stores, or home models—all use chlorofluorocarbons, the ozone-destroying gas. So, foods that require cold storage contribute to ozone depletion.

Packaging. The many layers of packaging are another matter, of course, and it's here that frozen foods lose out to fresh foods and most other packaged ones. The ideal packaging—none at all—won't work for frozen foods. Second best would be a package made of recycled and recyclable materials. This, too, has been elusive to frozen-foods makers.

What hasn't eluded them is the ability to heap layers of unrecyclable plastic upon unrecyclable coated paperboard to create massive amounts of household trash. A few products we examined had as many as six layers of packaging. In most cases, few of those layers were made of materials that were either recycled, recyclable, or reusable. Despite the needless trash these products generate, the food industry seems to love these products: they are frequently given food industry awards for innovation and design.

Microwave Hazards. Still another packaging issue are the health effects from microwave cooking some of these products in their packaging, as the instructions often recommend. A growing body of evidence suggests that some of the packages' chemicals "migrate" into the foods. (See "Zap!," page 39, for more on this.)

What to Look For

❏ Choose brands with the fewest layers of packaging. Ideally, there should be only one or two layers.

❏ Buy products that have reusable components, such as resealable containers or plastic dishes. But don't buy these if you don't plan to actually reuse these things. Note that some packages specifically warn, "Do Not Reuse."

❏ Be aware of packaging that states "Not to be used in a conventional oven," if that's where you plan to cook it.

[continued on page 82]

Frozen Appetizers

Best Choices

Chung King Eggrolls (Conagra Consumer Frozen Foods Co.)
Golden Potato Pancakes (Old Fashioned Kitchen Inc.)
Jeno's Pizza Rolls (Pillsbury Co.)
Oven Stuffs (Quaker Oats Co.)

Worst Choices

Matlaw's Egg Rolls with Lobster (Matlaw's Food Products Inc.)
Red L. Puff 'n' Puppies (Red L. Foods Inc.)
White Lotus Egg Rolls (Mui Li Wan Inc.)

Frozen Breakfast Foods

Best Choices

Aunt Jemima brand (Quaker Oats Co.)
Downyflake Waffles (Pet Inc./Whitman Corp.)
Eggo Waffles (Mrs. Smith's Frozen Foods Co.)
Morningstar Farms Cholesterol Free Patties (Worthington Foods Inc.)
Rich's Poly Rich Non Dairy Creamer (Rich Products Corp.)

Worst Choices

Cholesterol Free Egg Beaters (Nabisco Brands Inc.)
Great Starts Breakfasts (Campbell USA)
Swanson's Budget Breakfasts (Campbell USA)

Frozen Entrees

Best Choice

Budget Gourmet brand (AAGC/Bird's Eye) 

Worst Choices

Armour Classics (Conagra Consumer Frozen Foods Co.)
Campbell's Souper Combo (Campbell USA)
Healthy Choice (Conagra Consumer Frozen Foods Co.)
Stouffer's Dinner Supremes (Stouffer Foods Corp.)
Swanson products (Campbell USA)
White Castle Microwaveable Cheeseburgers (White Castle)

Fish Dishes

Best Choices

B.G. Brand Peeled Cooked Salad Shrimp (B.G. Shrimp Sales Co.)
Brilliant Shrimp Pops (Brilliant Seafood Co.)
Eat-All Stuffed Flounder (Eat All Frozen Food Co. Inc.)
Seapak Butterfly Shrimp (Rich-Seapack Corp.)
Taste O'Sea (Conagra Consumer Frozen Foods Co.)

Worst Choices

Booth Flounder Fillets (Booth Food Products)
Fullton's Ocean Fresh Fish (V & F Marketing Inc.)
Galletti Brothers Halibut Chunks (Galletti Brothers)
Seapak Salmon Steak with Seasoning Mix (Rich-Seapack Corp.)
Van De Kamp Natural Fillets (Van De Kamp's Frozen Foods)
Van De Kamp Microwave Fishsticks (Van De Kamp's Frozen Foods)

[continued from page 79]

The Bottom Line

We examined frozen packages on the market and found no clear winners. The best packaging came from Budget Gourmet, which managed to place its meals into a single layer of coated cardboard, but still neither recycled nor recyclable. Others had multiple layers, some made of uncoated cardboard or aluminum, both recyclable. Still others had sturdy plastic dishes that were at least reusable, assuming you were given to using plastic plates; perhaps they would make a good set of picnic dishes. But

Other Frozen Foods

Best Choices

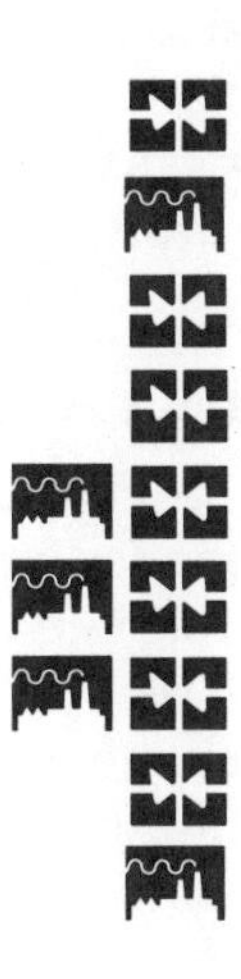

Bird's Eye brand (AAGC/Bird's Eye)
Celeste Pizza (Quaker Oats Co.)
Fox Deluxe Pizza (Pillsbury Co.)
McCain Elio's Pizzas (McCain Elio's Foods Inc.)
Moore's Onion Rings (Moore's Food Products)
Mrs. Smith's Pies (Mrs. Smith's Frozen Foods Co.)
Ore-Ida brand (Ore-Ida Foods Co.)
Stouffer's Pot Pies (Stouffer Foods Corp.)
Weight Watcher's brand (H.J. Heinz Co.)

Worst Choices

Kosher Empire Poultry Chicken Cut Up (Kosher Empire Poultry Inc.)
Lean Cuisine French Bread Pizza (Stouffer Foods Corp.)
Pepperidge Farms Croissant Crust Pizza (Pepperidge Farms Inc.)
Pillsbury Microwave Cheese Pizza (Pillsbury Co.)

many products featured materials neither recycled, recyclable, nor reusable: plastic films covering flimsy plastic bowls or trays, polystyrene containers, and coated cardboard.

The real bottom line is this: Avoid frozen entrees if possible. They are expensive, albeit handy, and generate a lot of waste. However, if you must buy one, choose a brand with the fewest layers of packaging, and with the most percentage of recyclable or reusable materials.

Meat, Poultry, and Fish

The first question Green Consumers should ask is: To eat or not to eat meat. That is the question a growing number of Americans are asking. It's not that vegetarianism is sharply on the rise—it has slowly grown to include about 7.5 million vegetarian Americans. But even meat eaters are eating less meat. Moreover, there are dramatic shifts in the kinds of meat people do eat.

What's the Problem?

Beef, heavily advertised on TV and in magazines as "real" food, is on the decline. That's good. Raising most beef cattle is a highly polluting and energy-intensive endeavor. We've already mentioned the vast amount of water it takes to produce the meat for just four quarter-pound hamburgers—2,500 gallons, according to John Robbins, author of *Diet for a New America*. (All told, raising livestock consumes more than half of all the water used in the United States.) Producing a pound of beef also requires the energy equivalent of one gallon of gasoline. But water and energy use are only the beginning. Consider this:

❑ Producing a pound of beef requires feeding cattle about 16 pounds of grain and soybeans. If Americans reduced their beef intake by just 10 percent, the amount of grain and soybeans saved could adequately feed 60 million people. That's roughly the number of people in the world who starve to death each year.
❑ More than 200 million acres of land have been deforested in the United States solely for the purpose of raising livestock. About one-third of all of North America is devoted to grazing. Some of the meat we eat comes from other countries. In Central America, for example, fully half the forests have been cleared for beef production. A lot of this land are the tropical rain forests. (More about that in a moment.)
❑ It takes about one-twentieth as much raw materials to grow grains and produce as it does to produce meat. In fact, the value of raw materials consumed in the United States to produce food from livestock is greater than the value of all oil, gas, and coal Americans consume.

So much for "real" food. Raising poultry minimizes some of these problems—for example, most commercial chickens don't need any grazing area because they are confined to coops—but it still requires considerable resources, including grain and water. According to one statistic, it takes about 400 gallons of water to produce a single serving of chicken, about one-sixth the amount needed for a hamburger patty.

Fish—and fish farming, a fast-growing source of Americans' seafood—is the least energy- and resource-intensive, although it is far from pollution-free; current farming techniques require dumping massive amounts of fish waste into waterways, which can affect other fish, plants, and organisms.

Beef and Rainforests. As we said, some beef eaten by Americans comes from tropical rain forests. According to the Rainforest Action Network (RAN), an environmental group, "The typical four-ounce hamburger patty represents about 55 square feet of tropical forest—a space that would statistically contain one 60-foot-tall tree; 50 saplings and seedlings representing 20 to 30 different tree species; two pounds of insects representing thousands of individuals and more than a hundred different species; a pound of mosses, fungi, and micro-organisms; and a section of the feeding zone of dozens of birds, reptiles, and mammals, some of them extremely rare. Millions of individuals and thousands of species of plant and animals inhabit a patch of tropical forest destroyed for a single hamburger."

Who uses the beef from rain forests? You won't find much of it in the supermarket as ground beef or steak; beef raised in rain forest regions is said to be stringy, tough, and cheap and often goes into mass-produced foods, where it is combined with fattier domestic beef and cereal products. RAN has urged Americans to avoid purchasing such processed-beef products as:

- ❑ baby foods
- ❑ canned beef products
- ❑ frozen beef products
- ❑ hot dogs
- ❑ luncheon meats
- ❑ soups

Some companies, such as Campbell Soup Company, have stated that while they import beef from South America, none of it comes from rain forest areas. But RAN responds that, regardless of whether this is true, there are other reasons to avoid all beef imported from Central and South America: "This beef is more likely to be contaminated with toxic chemicals, trace metals, and organic contaminants than that raised in the United States. Excessive pesticide residues have repeatedly been found in beef prepared by packing plants in Costa Rica, El Salvador, Guatemala, and Mexico, for which these packers have been decertified by the U.S. Department of Agriculture."

What about fast-food hamburgers? The major chains, such as McDonald's, Burger King, and Wendy's, all deny that they use rain forest beef; McDonald's has an unambiguous company policy that states that it, "does not, has not, and will not permit the destruction of tropical rain forests for our beef supply," and that "Any McDonald's supplier who is found to deviate from this policy or who cannot prove compliance with it will be immediately discontinued."

Organic Meat. The idea of meat being organic may seem contradictory at first, but it is growing in popularity. A relatively small number of organic ranchers are producing beef and poultry products that are free from the myriad of ingredients—antibiotics, pesticides, fertilizers, hormones, and other goodies—that are injected into or otherwise fed to livestock. Such techniques don't necessarily make organic meat "green"—relative to other foods, there are still significant impacts on the environment—but they are certainly "greener" than their nonorganic counterparts.

For more information about organic meat companies, contact the International Alliance for Sustainable Agriculture, 1701 University Ave. SE, Minneapolis, MN 55414; 612-331-1099.

Tuna and Dolphins. Ten days before Earth Day 1990, the three leading U.S. tuna canners made a dramatic announcement: They would stop buying tuna that were caught in a manner that harmed dolphins. It was a bold move, one heralded by environmentalists as a major step in stemming the unnecessary deaths of as many as 150,000 dolphins a year.

For reasons unknown, tuna often gather just below herds

of dolphin. So tuna fleets watch for dolphins to locate their catch. Then, in a practice known as "setting on dolphins," tuna fishermen snare the tuna—and dolphins—in plastic driftnets up to 30 miles long. Trapped, the air-breathing dolphins suffocate or drown. The dead or wounded dolphins are cast back into the sea.

The announcement—by H. J. Heinz Company, which markets StarKist; Van Camp Seafood Company, which markets Chicken of the Sea; and Bumble Bee Seafoods Inc.—was supposed to put an end to all this. And to a great extent, it did—but not entirely. By early 1991, there was still controversy over one brand, Bumble Bee, about whether it had made good on its promise. According to the Earth Island Institute, an environmental group, Bumble Bee was still using tuna processed in Thailand that were caught using driftnets. Bumble Bee claims that it was simply fulfilling old contracts, and that all its tuna will eventually be "dolphin-safe." As this book went to press, the matter remained unresolved.

The same cannot be said for other tuna and tuna products. StarKist, Chicken of the Sea, and Bumble Bee represent only 70 percent of the canned tuna market; the remainder are lesser-known brands, such as Geisha, Three Diamonds, and Carnation, and supermarket house brands, such as Safeway's Sea Trader label. Most of these are not dolphin-safe, according to the Earth Island Institute. It has sent letters to heads of the major supermarket chains, instructing them on how to determine the nature of the tuna they are buying.

Pet foods are another concern. Heinz's brands, including 9 Lives, have been deemed dolphin-safe, but no others have yet to pass Earth Island Institute's standards. Like supermarket house brands, much of the pet-food tuna is caught and processed in Thailand, making verification difficult.

Determining what's really dolphin-safe will be made easier with passage of two bills introduced in Congress. They would label cans either "dolphin safe" or state that the contents "were caught by methods which kill dolphins." Voters should write congressional representatives in support of such legislation.

For additional information about tuna and dolphins, contact the Earth Island Institute, 300 Broadway, Suite 28, San Francisco, CA 94133; 415-788-3666.

Packaging. As usual, packaging of meat and fish products is a concern. At the supermarket meat counter, the butcher paper of yesterday has given way to polystyrene plastic trays covered with polyethylene plastic wrap. Supermarket managers say this is necessary because, unlike the butcher shop of old times, today's customers can't inspect each cut of meat personally before it is wrapped. And plastic, they say, helps maintains meat freshness and provides sanitation on supermarket shelves.

Perhaps. But a growing number of supermarkets are responding to Green Consumer concerns by wrapping meat in paper for customers who request it. (See "The Green Supermarket," page 151, for names of some stores who do this.) Such policies may not be announced or posted in meat departments, so you'll have to ask.

As for packaged meat and meat products, the usual rules apply: the least amount of packaging, and the greatest percentage of recycled and recyclable packaging.

What to Look For

❑ Buy meat, poultry, and fish wrapped in paper instead of plastic whenever possible. Ask the store's fish or meat department to make this available to all customers.
❑ Avoid processed meat products. Some contain rain forest beef. Others contain low-grade beef from Latin America that contains pesticides and other undesirable ingredients.
❑ When buying packaged meats, look for minimally packaged products. One layer of packaging should be sufficient.
❑ Look for dolphin-safe brands of tuna. Make sure it states this specifically on the label.
❑ When shopping for tuna, buy the biggest-size can possible. Avoid convenience-size packages containing three single-serving cans in a cardboard container.
❑ Choose products from companies with good environmental records.

The Bottom Line

We found few outstandingly good or bad products in this category. Given that fresh meat, poultry, and fish vary from store to store, our look at packaged products revealed that the products tend to be packaged pretty much the same way, and that most packaging was appropriate for the product. Nearly all hot dogs, for example, come with one layer of clear plastic wrap with a paper label. In addition, our data base found no companies in this category that earned a top or bottom rating for environmental responsibility.

Luncheon meats, however, proved to be an exception when it came to packaging. In the name of convenience, some companies have opted for multiple layers of packaging, sometimes compartmentalizing different meats and other items. Probably the worst offender—and certainly a candidate for one of the dozen worst packaged products on the market—is Oscar Mayer Lunchables. They feature a luncheon meat (bologna, ham, etc.), crackers, and cheese, along with such accoutrements as a napkin, mustard container, and plastic knife for spreading. This is a needless waste, not to mention an overpriced item, given the amount of food contained within all that packaging.

Our advice: Stay away from packaged meats and meat products as much as possible.

Best Choices

*Bumble Bee Tuna (Van Camp Seafood Co.)
Chicken of the Sea Tuna (Ralston Purina Co.)
Deep Sea (Humble Whole Foods)
Fisherman's Net Sardines (L. Ray Packing Co.)
Ocean Light (Humble Whole Foods)
StarKist Tuna (H. J. Heinz Co.)

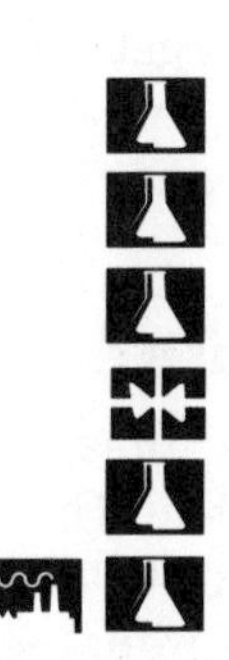

* As this book went to press, the matter over whether all of Bumble Bee's tuna was "dolphin-friendly" remained unresolved by Earth Island Institute. The company claimed that it was not selling any tuna in the United States that was caught using driftnet fishing methods. It is important to read each can's label for the "dolphin-safe" warranty.

Worst Choices

Beach Cliff Sardines (Stinson Canning Co.)
Beef Corn Dogs (Hillshire Farm & Kahn Co.)
Brunswick Kippered Snacks (Conners Brothers Ltd.)
Lloyd's Boneless Pork Chops (Lloyd's Food Products Inc.)
Oscar Mayer Lunchables (Oscar Mayer Foods Inc.)
Swanson Premium Chunk White Tuna (Campbell USA)
Underwood Sardines (Pet Inc.)

Fruits and Vegetables

We've already briefly discussed some of the environmental impacts of modern farming. (See "How Our Purchases Affect the Environment," page 16, and "Pesticides and Poisons," page 46.) The fact is, we've become a nation of consumers addicted to produce perfection: Perfectly shaped and blemish-free fruits and vegetables, available all year long. That's a tall order, to be sure, and American food growers, processors, and shippers have done an admirable job of catering to our every whim.

What's the Problem?

Want some juicy red tomatoes in the middle of February? No problem. They may be shipped to you from the fields of Puerto Rico or have been sailed, flown, and trucked several thousand miles from the farm to your family's dinner table. And all that time they will be kept under refrigeration to maintain freshness. Even domestic produce can travel great distances. The average mouthful of food travels about 1,300 miles to get to your dinner table, according to the Worldwatch Institute.

Of course, a lot of fruits and vegetables don't make the entire trip in their original form. Along the way, they are canned, bottled, frozen, or otherwise processed and packaged. For example, nearly half of all the fruits Americans consume go through some processing and packaging before being purchased and consumed.

Each step of the food-distribution process—growing, processing, transporting, warehousing, and selling—requires energy and resources, and contributes to pollution. Much of these resources are necessary to getting adequate and healthful food to our tables. But some of it is not. No one is suggesting that we go back to a diet composed only of locally grown and seasonal foods, but our developing tastes for convenience and exotic foods does clearly have important implications. In our environmentally conscious world, we may need to ask ourselves whether it really makes sense to bring so much of our food from so far away.

Packaging. A trip through the produce aisles yields some surprises, not just in the fruits and vegetables you'll find, but in the way some of them appear. Does a grapefruit, with its thick rind, really need to be shrink-wrapped in plastic? Certainly not, but you can find grapefruits—not to mention tomatoes, bell peppers, cauliflowers, lettuce, even bananas—constrained inside plastic wrappers, sometimes mounted on cardboard or polystyrene trays! There's simply no need for this.

And then there's the packaging for frozen, canned, and processed produce. Fortunately, the overwhelming majority of it is packaged in steel cans and glass bottles. But some of it comes in plastic, usually multi-resin plastics that cannot be recycled.

What to Look For

❑ Choose fresh produce whenever possible. Besides being healthier than the processed kind, it has the lowest "energy content" because it has used a minimal amount of resources to get it from farm field to dinner plate.

❑ Choose produce grown locally whenever possible. Ask the produce manager where you shop which fruits and vegetables come from nearby farms and which come from overseas. Make it known that you intend to avoid imported produce as much as possible.

❑ When selecting fresh produce, everything needn't be placed in its own plastic produce bag before being purchased. Several varieties of produce purchased by item (as opposed to by weight) can be "ganged" into one bag. At the checkout stand, the clerk will have no problem determining the price of each item.

❑ Ask for organic or no-pesticide produce. Besides being local, it requires fewer fertilizers and pesticides to grow. That's not only healthier for you, it's better for the environment.

❑ When buying prepackaged fruits and vegetables, choose products packaged simply, in aluminum, steel, and glass. Avoid plastic packages.

❑ Consider growing your own produce. Even if you have just a small plot of land, you can grow delicious tomatoes, peppers, and other produce. It will be healthful, economical, and fun.

❑ Choose products from companies with good environmental records.

❑ See also "How to Tell If It's Organic," page 50, for more information on buying produce.

The Bottom Line

Clearly, it was not possible for us to rate fresh produce in supermarkets nationwide. We'll leave it to you to follow the above suggestions wherever you shop. In examining prepackaged and processed fruits and vegetables, we found that most were packaged simply and well. Two major companies—Del Monte USA and Dole Food Co.—however, rated poorly for environmental performance; there were also two smaller companies with good ratings.

Best Choices

Barbara's brand products (Barbara's Bakery Inc.)

Eden brand products (Eden Foods Inc.)

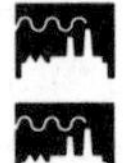

Worst Choices

Del Monte brand products (Del Monte USA)
Dole brand products (Dole Food Co.)

SNACKS AND SINGLE SERVINGS

Most of us were taught not to snack between meals, but as we've grown, times have changed. Just about any time is snack time, it seems, even mealtime. Indeed, for a lot of busy folks, meals—lunch in particular—consist of a series of snacks: a granola bar, some chips or crackers, a yogurt, candy bar, popcorn, pudding, or whatever.

Sometimes, a quick meal is in order. Increasingly, many of those meals seem to be microwaveable or ready-to-eat single-serving dishes. From salad and soup to main course and dessert, there is hardly an eat-on-the-go course you can't find on supermarket shelves.

Snacks and convenience are part of our way of life. Busy parents can have kids' meals standing by (kids can even make them themselves), brown baggers can get beyond the standard-issue tuna or bologna sandwich at lunch, and older and disabled people can prepare their own meals where they might otherwise rely on others—or simply skip eating altogether.

But convenience has a price, and that price can be high, both at the checkout stand and in the amount of energy consumed and trash produced through the manufacture, use, and disposal of these products.

Can we balance ease with environment? In most cases, the answer is yes, if you shop wisely.

Sometimes, a little creativity is in order. Want a microwaveable noodle soup? You can buy any of the half-dozen or so soups on the shelves, but all are packaged nearly the same: an unrecyclable plastic tub with a plastic or foil top, usually in a cardboard box, which may itself be shrink-wrapped.

Or, you could buy a package of ramen, a dry noodle product sold in most supermarkets. It comes in a single thin plastic wrapper. You'll have to use your own bowl and add water before microwave cooking, but if you've got a microwave to begin with, you've probably got access to these few simple resources. The difference between the two methods is not only a lot of trash. The overpackaged noodle soup sells for about 35 cents an ounce, compared to just 11 cents an ounce for the ramen.

There are many similar examples of ways to cut packaging and costs for snacks and convenient meals. It can be as simple as buying a large-size package, then doling out individual servings in resealable plastic containers.

What to Look For

❑ Choose snacks packaged in the least amount of packaging, and the highest percentage of recycled and recyclable materials. Especially avoid individually packaged single servings.
❑ Try to buy the largest-size package you can of a product. If you need smaller portions, divide them up yourselves in reusable and resealable plastic containers. Even cookies and chips can be divided into smaller resealable "zipper" plastic bags. (The bags can be rinsed out and reused.)
❑ When buying cardboard boxes of cookies and crackers, look for recycled boxes. (They have a gray underside; if it's white, it is unrecycled cardboard.)
❑ Choose products from companies with good environmental records.

The Bottom Line

We found a wide range of packaging among cookies, crackers, snacks, and single-serving products. Fortunately, in most categories, there were always one or more products packaged well. Sunshine Food Markets Inc. was one company whose products were often packaged simply in a single layer of recycled cardboard packaging.

We also found several companies whose environmental records rated at the top or bottom of all companies. Companies with good environmental records include the Quaker Oats Co.,

Eden Foods Inc., and H. J. Heinz. One of the most poorly rated snack food companies is also one of the biggest packaged food marketers: Nabisco Brands Inc. and the other subsidiaries of RJR/Nabisco: Del Monte USA, Dole Food Co., and Planters LifeSavers Co.

Cookies

Best Choices

Craquelin Puff Pastry (General Biscuit of America Inc.)
Estee Oatmeal Raisin Cookies (Estee Corp.)
Grahamy Bears (Sunshine Food Markets Inc.)
Lemon Coolers (Sunshine Food Markets Inc.)
Mini Middles (Keebler Co.)
O. T. Bears (Sunshine Food Markets Inc.)
Schoolhouse Cookies (Sunshine Food Markets Inc.)
Vanilla Wafers (Sunshine Food Markets Inc.)

Worst Choices

Barnum's Animal Crackers (Nabisco Brands Inc.)
Chips Ahoy (Nabisco Brands Inc.)
Cookie Break (Nabisco Brands Inc.)
Estee Original Sandwich Cookies (Estee Corp.)
Fig Newtons (Nabisco Brands Inc.)
Giggles (Nabisco Brands Inc.)
Honey Maid Graham Bites (Nabisco Brands Inc.)

[continued]

Worst Choices (Cont.)

Lorna Doone's Shortbread Cookies (Nabisco Brands Inc.)
Marshmallow Twirls Fudgecakes (Nabisco Brands Inc.)
Mrs. O'Days Delightful Cookies (O-D Associates Inc.)
Mystic Mints (Nabisco Brands Inc.)
Nilla Wafers (Nabisco Brands Inc.)
Oreo's (Nabisco Brands Inc.)
Peanut Butter Sandwich Cookie (Integrity Baking Co.)
Pepperidge Farm Cookies (Pepperidge Farm Inc.)
R.W. Frookies (R.W. Frookies Inc.)
Ritz Crackers (Nabisco Brands Inc.)
Teddy Grahams (Nabisco Brands Inc.)
Westbrae Natural Cookies (Westbrae Natural)

Crackers

Best Choices

American Heritage Crackers (Sunshine Food Markets Inc.)
Angonoa's Mini Bread Sticks
Breton Light Thin Wheat Crackers (Dare Foods Ltd.)
Butter Popped Corn Cakes (Quaker Oats Co.)
Club Crackers (Keebler Co.)
Estee Wheat Wafers (Estee Corp.)

Health Valley Graham Crackers (Health Valley Food Inc.)
100% Stone Ground Whole of the Wheat Cracker (Ak-Mak Bakeries)
Ryvita Toasted Sesame Rye (Shaffer & Clarke & Co.)
Stoned Wheat Thins (Red Oak Farms)
Townhouse Classic Crackers (Keebler Co.)
Wasa Crisp Bread (Sandoz Nutrition Corp.)
Weight Watchers Crispbread (H.J. Heinz Co.)
Wheat Cakes (Quaker Oats Co.)
White Cheddar Popped Corn Cakes (Quaker Oats Co.)

Worst Choices

American Classic Crackers (Nabisco Brands Inc.)
Better Cheddars (Nabisco Brands Inc.)
Bremner's Wafers (Bremner Inc.)
Cheese Nips (Nabisco Brands Inc.)
Chicken 'n' a Bisquit (Nabisco Brands Inc.)
Crispy Originals (Sunshine Food Markets Inc.)
Dandy Soup & Oyster Crackers (Nabisco Brands Inc.)
Devonsheer Rounds (Devonsheer)
Devonsheer Toasts (Devonsheer)
Do Dads (Nabisco Brands Inc.)
Escort Crackers (Nabisco Brands Inc.)
Hain's Crackers (Hain Pure Food Co.)
Harvest Crisps (Nabisco Brands Inc.)

[continued]

Worst Choices (Cont.)

Hi Ho's (Nabisco Brands Inc.)
Oat Thins (Nabisco Brands Inc.)
Ritz Crackers (Nabisco Brands Inc.)
Royal Lunch Milk Crackers (Nabisco Brands Inc.)
Sociables (Nabisco Brands Inc.)
Stoned Wheat Thin Crackers (Quail Oval Farms)
Swiss Cheese (Nabisco Brands Inc.)
Triscuits (Nabisco Brands Inc.)
Vegetable Thins (Nabisco Brands Inc.)
Wheat Thins (Nabisco Brands Inc.)
Wheatsworth (Nabisco Brands Inc.)
Whole Wheat Premium Plus (Nabisco Brands Inc.)

Dried Fruit

Best Choices

Fruit Leather (Stretch Allen Fruit Inc.)
Nature's Favorite Apple Chips (M.C. Snack Inc.)
Sun Maid Dried Fruit (Sun-Diamond Growers of California)
Weight Watchers Great Snackers (H.J. Heinz Co.)

Worst Choices

Del Monte products (Del Monte USA)
Dole products (Dole Food Co.)
Dromedary Pitted Dates (Fleischmanns Yeast Inc.)
Sun Maid Raisins (Sun-Diamond Growers of California)
Sunsweet Raisins (Sun-Diamond Growers of California)

Other Snacks

Best Choices

Eden Sea Vegetable Chips (Eden Foods Inc.)
Fisher Nuts (Fisher Nut Co.)
Franklin Crunch 'n' Munch Popcorn (American Home Foods)
Jiffy Pop Popcorn (American Home Foods)
Mauna Loa Macadamia Nuts (Mauna Loa Macadamia Nut Corp.)
Minute Tapioca (Kraft General Foods Inc.)
Mother's Rice Cakes (Quaker Oats Co.)
Orville Reddenbacher's Original Popping Corn (Beatrice Cheese Inc.)
Quaker Chewy Granola Bars (Quaker Oats Co.)
Rice Cakes (Quaker Oats Co.)
Snyder's of Hanover Pretzels (Snyder's of Hanover Inc.)
Weight Watcher's snacks (H.J. Heinz)

Worst Choices

Del Monte Pudding Cups (Del Monte USA)

Featherweight Sweet Pretenders (Sandoz Nutrition Corp.)

Hunt's Snack Pack Pudding (Beatrice/Hunt-Wesson Inc.)

Planter's products (Planters LifeSavers Co.)

Pop Secret Microwave Popcorn (General Mills Inc.)

Other Groceries

So far, our tour through the food aisles has hit the product categories that present some of the greatest challenges for Green Consumers. But the rest of the store offers opportunities, too. Amidst all those jars, cans, and other packages are products whose packaging or companies should be sought out—or avoided—by Green Consumers.

Breakfast Cereals

It takes only a few minutes' stroll through the cereal aisle to realize that nearly all dry breakfast cereals are packaged alike: in a recycled paperboard box with a sealed plastic bag. From Crispix to Cap'n Crunch to Corn Flakes, you won't find many variations on that theme, with the possible exception of the occasional unrecycled paperboard box.

The same cannot be said of cereal companies, however. There are some big differences here. Three companies—Arrowhead Mills, Kellogg, and Quaker Oats—rate highly on environmental policies and practices. Another company, Nabisco Brands Inc., rates poorly. The other major cereal company, Post, a division of Philip Morris, has not yet been rated.

Hot cereals have some packaging differences. Avoid products that come in individually wrapped pouches. In most cases, you can find the same products packaged in bulk.

Best Choices

- Apple Jacks (Kellogg Co.)
- Arrowhead Mills Puffed Rice & Puffed Corn (Arrowhead Mills Inc.)
- Cap'n Crunch (Quaker Oats Co.)
- Cocoa Krispies (Kellogg Co.)
- Common Sense Oat Bran (Kellogg Co.)
- Corn Flakes (Kellogg Co.)
- Corn Pops (Kellogg Co.)
- Cracklin' Oat Bran (Kellogg Co.)
- Crispix (Kellogg Co.)
- Crunchy Bran (Quaker Oats Co.)
- Froot Loops (Kellogg Co.)
- Frosted Flakes (Kellogg Co.)
- Frosted Mini Wheats (Kellogg Co.)
- Fruitful Bran (Kellogg Co.)
- Fruity Marshmallow Krispies (Kellogg Co.)
- Heartwise (Kellogg Co.)
- Instant Grits (Quaker Oats Co.)
- Just Right (Kellogg Co.)
- Life (Quaker Oats Co.)
- Mueslix (Kellogg Co.)
- Nut & Honey Corn Flakes (Kellogg Co.)
- Nutri-Grain (Kellogg Co.)
- Oatbake (Kellogg Co.)
- Oh's (Quaker Oats Co.)

[continued]

Best Choices (Cont.)

Old Fashioned Quaker Oats 100% Natural (Quaker Oats Co.)
Product 19 (Kellogg Co.)
Puffed Rice (Quaker Oats Co.)
Quaker 100% Natural (Quaker Oats Co.)
Quaker Oat Bran (Quaker Oats Co.)
Quaker Oat Squares (Quaker Oats Co.)
Quick Grits (Quaker Oats Co.)
Raisin Bran (Kellogg Co.)
Rice Krispies (Kellogg Co.)
Shredded Wheat Squares (Kellogg Co.)
Special K (Kellogg Co.)

Worst Choices

Cream of Rice (Nabisco Brands Inc.)
Cream of Wheat (Nabisco Brands Inc.)
Frosted Wheat Squares (Nabisco Brands Inc.)
Fruit Wheats (Nabisco Brands Inc.)
100% Bran (Nabisco Brands Inc.)
Shredded Wheat (Nabisco Brands Inc.)

Condiments and Preserves

It doesn't seem that long ago that pantry products were pretty much generic. There was "mustard," "ketchup," "mayonnaise," and so on. The differences were mostly in the brands and sizes of the products.

Things have changed a lot, not the least of which are our palates, some of which have become more sophisticated. Mustard, for example, comes as dijon, spicy, horseradish, light, no-salt, brown, and even natural stone-ground. Suddenly, shopping has gotten complicated.

Packaging has gotten more complicated too, albeit less so that on other supermarket aisles. Glass bottles and jars still prevail here, although plastics have begun to creep in. And not just any plastics. These are of the most undesirable variety, environmentally.

Consider squeezable ketchup containers, for example. They are made from seven layers of plastics and adhesives. These multiple layers make them unrecyclable. (An innovation, announced in 1990 by H. J. Heinz but not yet on the market as of this writing, would reduce the number of layers to five, and would use only one type of plastic, PET, increasing the chances that it can be recycled. But even that package won't be recyclable by all communities that accept PET plastic.) And while squeezable ketchup containers do solve one of the great dinner table dilemmas—getting the ketchup out of the bottle gracefully—they are far from environmentally desirable.

But all hope is not lost. If you like to squeeze your ketchup instead of pour it, fine. Buy one of these containers. But when it is empty, don't throw it away. Buy the biggest glass bottle of ketchup you can find and refill the plastic container. So, you can have your ketchup and squeeze it, too!

It is easy to identify these multiple-material containers. They feature the number "7" inside a recycling triangle on the container's bottom. If the product has no number at all stamped on the bottom, you can safely assume that it is a multimaterial container and is therefore unrecyclable; it is unclear how Heinz's improved squeezable bottle will be classified. (See "Which Plastic Is It?," page 35, for more on plastic package codes.) In any case, opt for glass bottles whenever you have a choice.

Best Choices

Anything packaged in glass.
Anne's Original Salad Dressings (Anne Lanyi Foods)
Arrowhead Mills Peanut Butter (Arrowhead Mills Inc.)
Eden products (Eden Foods Inc.)
Heinz products (H.J. Heinz Co.)
Hidden Valley dressings (Kingsford Products Co.)
K.C. Masterpiece Mesquite Barbecue Sauce (Kingsford Products Co.)
Natural Chocolate Sauce (Barbara's Bakery Inc.)
Newman's Own products (Newman's Own Inc.)
Salad Crispins (H.V.R. Co.)
Smucker's products (glass packaging) (J.M. Smucker Co.)
Uncle Chang's Sesame Seed Dressing (Anne Lanyi Foods)
Weight Watchers products (H.J. Heinz Co.)

Worst Choices

Any squeezable plastic container.
Brownberry Croutons (Best Foods)
Cool Whip (AAGC/Bird's Eye)
Del Monte brand (Del Monte USA)
Dutch Gold Honey (Dutch Gold Honey Inc.)
Estee Creamy French Style Dressing (Estee Corp.)
French's America's Favorite Mustard (squeezable) (R.T. French Co.)

[continued]

Grey Poupon Mustards (Nabisco Brands Inc.)
Grower's Company Selection Nouvelle Garnies (Millflow Spice Corp.)
Heinz Ketchup (squeezable bottle) (H.J. Heinz Co.)
Hunt's Squeezable Plastic Tomato Catsup (Beat./Hunt-Wesson Inc.)
Jif Peanut Butter (Procter & Gamble Co.)
Knorr Dry Mixes (Special Products)
Kraft Squeeze Real Mayonnaise (Kraft General Foods Inc.)
La Creme Whipped Topping (Pet Inc.)
McCormick Bac'n Pieces Imitation Bacon Chips (McCormick & Co.)
Ocean Spray Jellied Cranberry Sauce (Ocean Spray Cranberries Inc.)
Ortega Salsa (Nabisco Brands Inc.)
Peter Pan Peanut Butter (Beatrice/Hunt-Wesson)
Planter's products (Planters LifeSavers Co.)
Sue Bee honeys (Sioux Honey Co-Op)
Utz Dips (Thomas J. Lipton Co.)
Vermont Maid Syrup (Nabisco Brands Inc.)
Welch's Squeezable Concord Grape Jelly (Welch Foods Inc.)
Wishbone dressings (Thomas J. Lipton Co.)

Cooking and Baking Needs

Much of what we said about condiments and preserves also holds true for this category. Plastic, once again, seems to be gaining in popularity—at least among manufacturers. Particularly troublesome are a new breed of pancake mixes that feature a half-full plastic bottle of pancake batter mix. You need only add water and shake—*voila*, pancake batter. However convenient, we find the idea of a half-filled plastic bottle of anything particularly offensive. Here's a tip: Buy one full container (preferably, packaged in a cardboard box), pour premeasured amounts into your own containers (perhaps reusing one of the commercial brands). Put them in the refrigerator. When you want pancakes or waffles in a hurry, take one out, add water, and shake. *Voila*, pancake batter. In other words, do it yourself. You'll save both trash and cash.

Another product to avoid is aerosol spray cooking oils. There is usually little an aerosol spray can do that careful application by other means cannot accomplish. As we've stated, aerosols are not readily recyclable. Moreover, you get far less product per dollar than when buying nonaerosol versions of the same products.

Best Choices

Any cooking oil in a glass bottle
Argo Corn Starch (Best Foods)
Arm & Hammer Baking Soda (Church & Dwight Co.)
Arrowhead Mills Blue Corn Pancake Mix (Arrow. Mills Inc.)
Arrowhead Mills flours and grains (Arrowhead Mills Inc.)
Aunt Jemima Pancake Mix (boxed) (Quaker Oats Co.)
Bisquick (boxed) (General Mills Inc.)
Cream Pure Corn Starch (Dial Corp.)
Flako Corn Muffin Mix (Quaker Oats Co.)
Hershey's Baking Chocolate (Hershey Foods Corp.)
Jell-O Gelatin Dessert (Kraft General Foods Inc.)
Knorr Milk Chocolate Mousse Dessert Mix (Special Products)
Lawry's Season Salt (Lawry's Foods Inc.)
Mrs. Butterworth's Buttermilk Pancake Mix (Van Den Bergh Foods)
Pillsbury Hungry Jack Pancake Mix (Pillsbury Co.)

Pillsbury Microwave cake mixes (Pillsbury Co.)

Quaker Yellow Corn Meal (Quaker Oats Co.)

Sure Gel Fruit Pectin (Kraft General Foods Inc.)

Worst Choices

Any cooking oil in a plastic bottle

Aunt Jemima Pancake Express Mix (plastic) (Quaker Oats Co.)

Baker's Joy aerosol oil (Alberto Culver Co.)

Betty Crocker Creamy Deluxe Frostings (General Mills Inc.)

Betty Crocker Micro Rave Cake Mix (General Mills Inc.)

Bisquick Shake 'n' Pour Pancake Mix (General Mills Inc.)

Cake Mate Decorating Icing (McCormick & Co. Inc.)

Crisco Pure Vegetable Oil (Procter & Gamble)

Fleischmann's Yeast (Fleischmann's Yeast Inc.)

Nabisco Graham Cracker Crumbs (Nabisco Brands Inc.)

Mazola No-Stick Cooking Spray (CPC International Inc.)

Pam No-Stick Cooking Spray (Boyle-Midway Household Prod. Inc.)

Pepperidge Farm Pancake Mix (Pepperidge Farm Inc.)

Pillsbury icings (Pillsbury Co.)

Planter's Peanut Oil (Planters LifeSavers Co.)

Reese Swedish Pancake Mix (Reese Finer Foods)

Reynolds Baker's Choice Bake Cups (Reynolds Metals Co.)

Weight Watchers Buttery Spray (Ore-Ida Foods Co. Inc.)

Pastas, Beans, Grains, and Soups

Pastas and beans represent some of the greenest foods, given the protein and other nutrients they provide compared with the amount of energy and other resources required to grow them. We found most product categories similarly packaged: pastas were usually found in recycled cardboard boxes, beans in steel cans or plastic bags, soups in steel cans, etc.

All-in-one convenience products produced some disappointments, however. Products like Minute Microwave Dishes, a microwaveable meal, were heavily overpackaged. A typical version features two multimaterial pouches contained in a plastic microwaveable dish with a plastic cover, which in turn is nestled in a paperboard box, shrink-wrapped with yet more plastic. Clearly, this is a wasteful item (and overpriced: the Minute macaroni and cheese was three times more expensive than the nonmicrowave variety). Nearly all of the microwaveable and instant soup mixes were overpackaged and overpriced, compared to their nonmicrowave counterparts. (Many of those counterparts could be made in a microwave, assuming you used your own cooking dishes and followed the simple instructions that come with your microwave oven.)

As usual, there were some environmentally outstanding—good and bad—companies in this category.

Best Choices

Buitoni pastas (Buitoni Foods Corp)
Eden products (Eden Foods Inc.)
Fantastic Instant Refried Beans (Fantastic Foods Inc.)
Golden Grain pastas (Golden Grain Co.)
Heinz products (H.J. Heinz)
Light & Fluffy Egg Noodles (Hershey Foods Inc.)

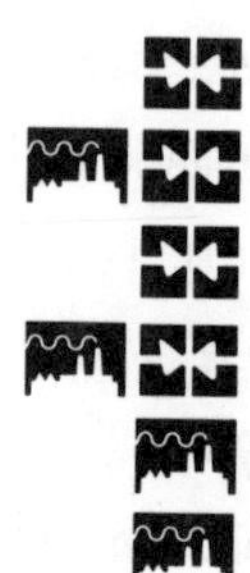

Maruchan Ramen Noodle Soup (Maruchan Inc.)
Nissan Oodles of Noodles soups (Nissin Foods)
Noodle Roni (Golden Grain Co.)
Quaker Barley (Quaker Oats Co.)
Reese rices (Reese Finer Foods)
Rice-A-Roni products (Golden Grain Co.)
San Giorgio Pastas (Hershey Foods Inc.)
Sanwa Ramen Pride (Sanwa Foods Inc.)
Weight Watcher soups and stews (H.J. Heinz Co.)

Worst Choices

Campbell's Microwave Soups (Campbell Soup Co.)
College Inn Chicken Broth (Nabisco Brands Inc.)
Cup O'Noodles Instant Soups (Nissan Foods)
Fantastic Foods Couscous (Fantastic Foods Inc.)
Fantastic Noodle Instant Soup (Fantastic Foods Inc.)
Featherweight Healthy Recipes (Sandoz Nutrition Corp.)
Knorr Risotto (CPC International Inc.)
Lunch Bucket Microwaveable Soups (Dial Corp.)
Minute Microwave Dishes (Kraft General Foods)
Pasta Valente (Pasta Valente)
Uncle Ben's Microwave Rice (Uncle Ben's Inc.)
Wick Fowler's Famous Family Style Chili Kit (Caliente Chile Inc.)

Air Fresheners

Life doesn't always smell as sweet as it should. Sometimes, bad odors invade our lives, whether created by people, animals, or mysterious things that go bump in the night. And when they do, many of us turn to a variety of commercial products, known variously as air fresheners or room deodorizers. They come in aerosol, liquid, gel, and solid form, usually accompanied by pictures of flowers and dubbed with a vague but highly suggestive name—Mountain Heather, Country Garden, Summer Breeze, Rainshower Fresh, or Early Spring, to cite just a few examples of the many air-freshening products found on our local supermarket's shelves.

We love our sweet-smelling air. Americans buy more than 300 million air fresheners annually to the tune of nearly a half-billion dollars, and sales are up 17 percent during the past year, according to Arbitron/SAMI, a product-tracking service.

But are air fresheners safe for our health and the environment? There is no doubt that there are kinder, gentler, and more effective remedies that cost a lot less than most commercial air-freshener products.

What's the Problem?

Finding out what is contained in an air freshener is no mean feat. Product labels rarely, if ever, state ingredients. This is proprietary information, secret as the recipe for Coca Cola. And despite the intended use of these products, neither the Environmental Protection Agency, the Food and Drug Administration, the Consumer Products Safety Commission, nor any other federal government agency keeps track of these products' ingredients. Suffice to say the companies themselves weren't overeager to help us ferret out this information.

Still, while the details may be a bit foggy, there is much that can be said to clear the air on this product.

Pollutants. In whatever form they are sold, most air fresheners contain anywhere from 20 to 90 percent volatile organic compounds, or VOCs. VOCs aren't necessarily bad; they occur

Clean Machines

One effective way to clear the air in your home or office is by using indoor plants. That's right, those philodendrons, chrysanthemums, and other plants are increasingly being recognized as natural air cleaners. Indeed, according to an impressive federal study, a few everyday potted plants can become outright life savers.

The principal basis of this finding is research conducted by the National Aeronautics and Space Administration (NASA), which is interested in ways to reduce pollutants inside future space habitats. In such sealed environments, humans quickly pollute the air with gases produced from breathing, and small amounts of gases given off by furnishings and equipment further complicate matters.

A two-year study by NASA found that plants can remove up to 87 percent of toxic indoor air within 24 hours, including such dangerous pollutants as formaldehyde, benzene, and trichloroethylene. NASA researchers concluded that plants "have demonstrated the potential for improving indoor air quality by removing trace organic pollutants from the air."

Some indoor house plants, NASA found, have the ability to absorb many air pollutants through their leaves and roots and convert the pollutants into breathable oxygen. Thus, indoor air is cleansed by the same process nature uses outdoors.

What Works With What

According to NASA, it takes about one plant for every 100 square feet of space to provide effective cleaning. That's a lot of plants for most homes. For example, it would take 15 to 20 golden pothos and spider plants to clean and refresh the air in the average 1,800-square-foot home. Specific plants work best with specific pollutants. For example, English ivy is recommended to absorb tobacco smoke, Bamboo palm is recommended for carpet odors, and mother-in-law's tongue is recommended for household cleaner smells.

Besides homes, plants can provide air-cleaning benefits in offices and even in industrial environments. In some sophisticated approaches, building designs can build plant filtering systems into atriums, lobbies, and walkways. For example, dry cleaning fluid smells can be absorbed by gerbera daisy, paints by peace lily, and varnishes by warneckei. Indoor air pollution costs American businesses as much as $60 billion a year in sick leave, lost earnings, and lost productivity, according to the Foliage for Clean Air Council, an industry group. A handful of relatively inexpensive plants may help to lower those costs.

For more information about the indoor-air cleaning properties of plants, contact the Foliage for Clean Air Council at 405 N. Washington St., Falls Church, VA 22046; 703-534-5268.

naturally. Everything you can smell contains VOCs. It's their volatility that makes them smell.

But some VOCs are pollutants. And while the industry points out that the fragrances it uses are derived from substances found in nature—roses, for example—that does not necessarily make them good for the environment, according to Marco Kaltofen of the National Toxics Campaign. He explains that while the atmosphere has developed harmless ways to absorb the VOCs emitted by roses (and most other things in nature), this doesn't hold true when these same things are manufactured by humans. Moreover, many of the VOC propellants used in air fresheners, such as butane and propane, are petroleum compounds that contribute to the creation of low-level ozone, or photochemical smog.

Packaging. Packaging presents familiar problems. Aerosol cans are unrecyclable and generally undesirable, of course. And most air-fragrance products come in disposable plastic packages. One product, Glade Plug-Ins, from S.C. Johnson, requires electricity to operate, contributing to further wastefulness.

Still, the industry prefers to view its ingredients as "natural" and its products as essential. "Air fresheners are not a frivolous item," an industry spokesman told us. "They serve some very strong social benefits. People want to feel good about their surroundings."

Perhaps. But air fresheners tend to hide problems more than solve them, says John Banta, a home environmental specialist and owner of Baubiologie Hardware in Pacific Grove, California. "If you have to use an air freshener, there's something wrong. You're treating the symptom and not the problem." Banta suggests geting rid of bad odors with thorough cleaning, proper ventilation, and inspection for molds and spores, both of which can result from use of humidifiers. "Sometimes the problem is right there under your nose," he says.

The Bottom Line

From a Green Consumer perspective, we can't honestly recommend any of these products. Air fresheners are generally wasteful, and their contents may not be good for the environment or your health. In general, they tend to cover up smells more than get rid of them, and that can sometimes mask more serious problems: simple dirt and grime, water leaks and other plumbing problems, old moldy carpets, heating and air conditioning malfunctions. Persistent odors may require professional help.

Still, if you still would like to sweeten the air, try one of the home remedies below.

Doing It Yourself

Baking soda: This is the best and safest absorber of most household odors. Pour a few tablespoons in several open containers and place them throughout your home. They work best in small enclosed areas, such as pantries and closets.

Pomander: Stick as many cloves as possible into a whole orange and hang it from a string. Over time, the orange won't spoil; it will simply dry out to a golfball-like hardness. In the meantime, it will give off a delightful odor.

Potpourri: Place your favorite herbs and spices in a small basket or jar or in small sachet bags.

Spices: Boil cinnamon, cloves, and other sweet-smelling herbs in a cheesecloth bag.

Baby Products

You probably don't need to be a parent to know that babies can cost a small fortune—for a middle-income family, about $12,000 during the first two years of a baby's life, according to government statistics. That includes doctor bills, which is out of the scope of this book. But there are many other purchases where you can make "green" choices.

What's the Problem?

Perhaps no product has been debated on environmental terms quite as much as diapers has.

The debate is a Green Consumer classic: How to balance environmental responsibility with the convenience demanded (and usually deserved) by busy, working people.

With diapers, the debate has been exceptionally loud because the product is such a visible one, contributing about 2 percent to the nation's landfills—roughly 4 billion pounds of discarded diapers a year.

But it's not just the sheer volume or weight of the trash that's at issue. It's also what's contained within each of these individual bundles of paper and plastic. Estimates have it that more than 18 million diapers a year contain about 2.8 million tons of excrement and urine. An infant's feces and urine can contain any of over 100 viruses, including live polio and hepatitis from vaccine residues. These viruses are potentially hazardous to sanitation workers and to others, both through groundwater or when carried by flies.

Further, the materials themselves are not to be sneezed at, including some 67,500 tons of polyethylene resin a year, used in the waterproof liners of most commercial brands. When disposed of, that plastic takes between 300 and 500 years to break down. (Or so researchers say. No one really knows; these diapers have only been around for about twenty years.)

The alternative to disposables—cotton diapers—have their environmental drawbacks, too. Though reusable (a cloth diaper can be washed and reused up to 150 times), each washing requires water, detergent, and electricity. If the diapers are picked up and delivered by a commercial service, the energy and pollution of the delivery truck also must be factored in. Sanitation, however, is not considered a problem with cloth diapers, because home or commercial washing machines channel human waste into the sewage system where it can be properly removed from the water system.

Also in their favor, cloth diapers—even when using a diaper service—cost about one-third less than the same number of disposables. Considering that babies require between 6,000 and 10,000 diaper changes over a three-year period, that savings alone can run to $700.

Biodegradable Diapers. What about biodegradable diapers? When introduced a few years back, they seemed a good idea—what could be better than a diaper that simply disappeared? But as with other disposables, the hype didn't live up to reality. Biodegradables, it turns out, don't really degrade in landfills. In 1990, at least one company that heavily promoted them was fined and forced (by a task force of ten state attorneys general) to stop making biodegradability claims.

The New Cloth Diapers. Today's cloth diapers have been improved upon since the good old days. Velcro, that miraculous material, has made fastening and unfastening diapers easier, and there have been other innovations, too. A few diaper makers have tried to combine the best of both worlds. For example, one brand, Cottontails, offers a version of its form-fitted, cotton-knit diaper with a disposable plastic liner, combining reusability with minimal disposability. Unfortunately, these and many other cloth diapers aren't readily available on supermarket shelves. Diaper buyers should urge grocery stores to carry cloth brands. In the meantime, you'll have to look for cloth diapers in department stores, baby specialty shops, and mail-order catalogs. (See "Shopping by Mail," page 182, for a list of catalogs offering environmentally responsible products.)

Compostable Diapers? The two biggest disposable diaper makers—Procter & Gamble's Pampers and Luvs and Kimberly-Clark's Huggies control 80 percent of the market—have been actively trying to head off the anti-disposable movement with their own innovations. Procter & Gamble experimented with a disposable diaper pickup service in the Seattle area, which separated out and sanitized recyclable components of the diapers. But such programs beg the question: If parents are willing to separate their dirty disposable diapers from their other trash and have them picked up for recycling, why can't they just as easily use a cloth diaper service?

Perhaps more promising, Procter & Gamble has committed at least $20 million to support development of diaper-composting plants. It hopes to develop a 100 percent compostable diaper and a method for turning diapers into humus, an important ingredient of fertile soil; the humus could be used for landscaping or farming applications. The process might work: More than 150 U.S. cities have composting facilities in the planning stages.

The Bottom Line

In the end, choosing between disposable and cloth diapers is a highly personal decision. Green Consumerism aside, there is contradictory evidence as to the health effects of cloth versus disposables. Whereas some parents contend that disposables containing superabsorbent gel (Ultra Pampers is one such brand) keep babies drier, others swear that cloth diapers help prevent dermatitis—diaper rash.

Some decisions are circumstantial. For example, many day-care centers require that parents provide disposable diapers for their children. Other decisions are practical: when traveling away from home, disposables are a hands-down improvement over cloth.

But that doesn't mean disposables are needed all the time. Perhaps the best solution is a mixed one: to use cloth diapers as often as possible, substituting disposables when cloth isn't practical.

What about the differences among disposables? The products themselves are virtually indistinguishable. But the companies aren't. When examining their environmental records, the Council on Economic Priorities rated Procter & Gamble higher than its competitor, Kimberly-Clark.

Baby Foods

We once considered baby food as pure as the driven snow—select blends of apples, carrots, potatoes, and other fruits of the earth. Sadly, today's apples, carrots, potatoes, and other produce are no longer assumed to be either pure or safe. From concerns about residues of the pesticide Alar on apples to baby-food company disclosures of synthetic ingredients being sold as 100 percent natural ingredients, many parents have had second thoughts about the purity of some baby foods.

Consumer concerns about chemicals in food have led to the spawning of the organic baby food industry. Still a relative toddler, with less than 1 percent of the $1 billion-a-year baby food business, the organic baby food market is growing steadily as its products become more widely available. The market received a boost from a 1989 study by the Natural Resources Defense Council that warned of lifetime risks of cancer for today's schoolchildren from pesticide residues. According to the NRDC, between 5,500 and 6,200 of today's preschoolers are likely to develop cancer solely because of exposure to just eight pesticides.

The two leading organic baby food companies, Earth's Best and Simply Pure, both offer 100 percent certifiably organic products. Prices for both products are considerably higher than non-organic varieties (Gerber, Beech-Nut, and Heinz control more than 90 percent of the market), but prices are expected to decline as the market grows and other companies begin marketing to health-concerned parents. Both brands are available in select supermarkets, as well as in health food stores.

Keep in mind that making your own baby food is not difficult and is the most economical and ecological way to go. You will be sure to avoid additives and unnecessary packaging, and can personally ensure that your baby is getting a healthy, balanced diet.

Other Baby Products. Besides foods and diapers, there are plenty of other supermarket purchases required to help keep infants clean, healthy, and happy. Most soaps and shampoos are packaged in plastic, to be unbreakable; few of the plastics most commonly used are among the most recyclable ones. As usual, there are good companies and bad ones among baby-product manufacturers.

What to Look For

❑ Choose baby foods packaged in recyclable containers, such as glass jars and steel or aluminum cans.

❑ Buy the largest-size product you can. Avoid single-serving containers whenever possible.

❑ If you are concerned about pesticides, buy organic baby foods. Better yet, make your own.

❑ Use cloth diapers whenever possible. Use disposables only when absolutely necessary.

❑ Choose products from companies with good environmental records.

Best Choices

A & Z Ointment (Schering-Plough Corp.)
Aimentum Protein Hydrasolate Formula with Iron (Ross Labs)
Biobottoms Diaper Covers (Biobottoms)
Bumkins Cloth Diapers (Bumkins Family Products)
Cotton Cloud Diapers (Baby Bunz & Co.)
Earth's Best Baby Food (Earth's Best Inc.)
Fisher-Price products (S.C. Johnson & Son Inc.)
Gerber Oatmeal Cereal for Baby (Gerber Products Co.)
HappiNappi Diapers (Baby Bunz & Co.)
Heinz baby foods (H.J. Heinz Co.)
Isomill Soy Protein Formula (Ross Labs)
Johnson's baby products (Johnson & Johnson)
Pedialyte for Maintenance of Body Water and Minerals (Ross Labs)
Simply Pure Baby Food (Simply Pure Baby Food)

Worst Choices

Baby Fresh Hypoallergenic Swipes (Scott Paper Co.)
Huggies Disposable Diapers (Kimberly-Clark)
Luvs (Procter & Gamble Co.)
Pampers (Procter & Gamble Co.)
Wash A Bye Baby (Scott Paper Co.)
Zwieback Teething Toast (Nabisco Brands Inc.)

BATTERIES

Americans go through a lot of batteries—over two billion a year to power an estimated 900 million devices. And with all those colorful little cylinders and silver buttons come a battery of environmental hazards.

What's the Problem?

For starters, the dry-cell batteries that power radios, calculators, computers, watches, and other things contain a number of toxic substances, including cadmium, lithium, manganese dioxide, mercury, nickel, silver, and zinc. Though their amounts may be tiny, these materials can be problematic when they leak into the water supply after being disposed of in landfills, or when they end up in incinerator ash. Cadmium and mercury are linked to cancer and neurological disorders, for example, while manganese is associated with respiratory problems and zinc can endanger wildlife.

The battery industry maintains that it is acceptable, even preferable, to dispose of batteries in our normal household trash; they point out that a bucketful of batteries could discharge each other, creating the possibility of explosions. But others disagree. Two states, Connecticut and Minnesota, have introduced legislation that views batteries as hazardous waste, requiring that they be disposed of in more responsible ways.

Dangerous chemicals aren't batteries' only problem. Battery manufacturing requires a lot of energy resources. In fact, manufacturing a standard disposable household battery uses 50 times more power than the battery will generate over its lifetime.

Disposables vs. Rechargeables

Dry, or household-use batteries come in two types: rechargeable (nickel-cadmium), and disposable, of which alkaline is the most popular. Rechargeables comprise only about 7 or 8 percent of the battery market.

There are trade-offs among battery types. Disposables have a small amount of mercury in them to prevent a volatile chemical reaction between battery components. The amount of mercury in batteries has decreased considerably in recent years in response to concern about its effect on the environment. Carbon-zinc batteries, the cheapest on the market, contain less mercury but don't last as long as other types. Over time, you may use more batteries—and more mercury.

Nickel-cadmium batteries can be recharged up to 1,000 times. But their performance time is shorter than most disposables; many users keep two sets—one to recharge while the other is in use. About three-fourths of nickel cadmium batteries are permanently sealed in products and appliances, such as rechargeable phones, razors, drills, computers, and vacuums. The Connecticut and Minnesota bills would require such batteries to be removable so they can be recycled and replaced.

Button-cell batteries—used in hearing aids, calculators, and watches—are economical to recycle because they contain a high percentage of mercury. Silver is also a valuable recoverable resource. Currently, only button-cell batteries are being recycled in the United States. Check with a local hearing aid dealer or jewelry store to see if they collect button cells. Many do.

According to the Natural Resources Defense Council, there are technically and economically feasible systems for recovering cadmium, mercury, and silver from many common battery types. But recycling has not yet caught on in this country.

What to Look For

❑ Invest in a rechargeable battery "system." They are an ecological and economical solution. Though rechargeables are more expensive and require purchasing a recharger, if your family uses a lot of batteries, they can save you $100 a year or more.
❑ Send your used batteries back to the manufacturers by mail, signifying that you consider used batteries a potential danger. This may encourage companies to begin battery recycling programs for consumers.
❑ If you buy disposable batteries, dispose of them at a local hazardous waste collection site if possible. Some communities have regular pick-ups, others have drop-off stations. If there is no collection program in your area, consider purchasing mercury-free disposables. They are available from Power Plus of America (1605 Lakes Pkwy., Lawrenceville, GA 30243; 404-339-

1672), as well as some mail-order companies.

❑ In the end, ask yourself whether you really need products that require batteries. Purchasing a battery-run toy for your child contributes to his or her inheritance of solid waste problems.

The Bottom Line

The most environmental batteries—rechargeables—aren't readily available in most supermarkets, although this is destined to change as major battery companies such as EverReady and Rayovac introduce rechargeables and rechargers.

In the meantime, you can find rechargeables in hardware stores and from mail-order companies (see "Shopping by Mail," page 182). Other sources include:

❑ Gates Energy Products (P.O. Box 861, Gainesville, FL 32602; 904-462-3911) makes Millenium rechargeables.

❑ SAFT America (711 Industrial Blvd., Valdosta, GA 31601; 912-247-2331) makes Again and Again rechargeables.

❑ SunWatt Corporation (Box 751, Addison, ME 04606; 207-497-2204) sells the F-2A solar charger, which uses the sun's power. It takes 1 to 3 days to recharge six batteries, compared with six to eight hours for electric rechargers.

Household Cleaners

Let's start with the bad news: There is little that is clean or green about most commercial household cleaning products.

That holds true across the board, from Ajax and Fantastik to the newest so-called "biodegradables." The fact is, almost anything we dump down our drains, even if derived from "natural" substances, can cause environmental problems.

The first problem is that there are few concrete definitions or legal requirements for most of the common buzzwords used on product labels—words like "nontoxic," "biodegradable," and "natural." And manufacturers aren't required to disclose most of their ingredients—they are considered trade secrets—and even when disclosed, the list of hard-to-pronounce chemicals (example: *sodium lauryl ethoxysulfate*) mean little to most of us.

Equally confusing is the trade-off between environmental and human safety. Some environmentally preferable products are more harmful to people than their less-ecological alternatives. Low-phosphate detergents, for example, while reducing water pollution, are 100 to 1,000 times more caustic than phosphate detergents, according to the Household Hazardous Waste Project (HHWP). This means low-phosphate detergents can cause serious burns if even a small amount is ingested.

The good (or at least better) news is that there is an increasing number of greener products on the market, though comparing one to the others may require a degree in chemistry, to say the least. And while none of these warrants our unqualified recommendation, some offer improvements over their competitors.

What's the Problem?

Biodegradability. One of the popular claims among today's alternative cleaners is that they are "biodegradable." This implies that a cleaner's ingredients break down quickly and harmlessly after they go down the drain.

Biodegradability refers to the rate and thoroughness with which a cleaner's ingredients break down. Under traditional definitions, when something biodegrades, the end product should consist solely of carbon dioxide, water, and salts. Moreover, the process must take place quickly enough to avoid causing harm before the biodegradation is complete. Those ingredients that break down slowly or incompletely endanger plants, animals, and micro-organisms that live in water or soil systems.

According to experts, many so-called biodegradable cleaning products on the market contain ingredients that do not degrade fully or quickly enough to minimize their environmental impact. Even at low concentrations, some ingredients can increase the penetration of other harmful chemicals through the protective layers of plants and animals. When Green Cross scrutinized the biodegradability claims of 40 cleaning products, almost none met its standards.

Making Your Own

There's no real secret to making your own cleaning products. Indeed, doing so produces the cheapest, safest, and "greenest" products available. There are dozens of do-it-yourself recipes for household cleaners, many using common ingredients you probably already have: water, ammonia, baking soda, borax, and vinegar. One good source of recipes is *Clean & Green: The Complete Guide to Nontoxic and Environmentally Safe Housekeeping*, by Annie Berthold-Bond (Ceres Press, $8.95). The book contains nearly 500 recipes for cleaning everything from fabrics to Formica to floors.

Here are some other quick-and-easy solutions:

All-purpose cleaner: Mix two teaspoons of borax and one teaspoon of soap in a quart of water. Or, use a half cup of washing soda (hydrated sodium carbonate) in a bucket of water; it works on all but aluminum surfaces.

Automatic dishwashing liquid: Use equal parts borax and washing soda (hydrated sodium carbonate). Don't use if you have hard water.

Floor cleaner: No-wax linoleum can be washed with a half cup of white vinegar mixed with a half gallon of warm water.

Glass cleaner: Clean with a mixture of equal parts white vinegar and water.

Oven cleaner: Clean oven with a paste of baking soda, salt, and hot water. Or sprinkle with dry baking soda and scrub with a damp cloth after five minutes. Don't let the baking soda touch wires or heating elements.

Toilet cleaner: Use soap and borax; remove stubborn rings and lime buildup with white vinegar. Or try sprinkling baking soda into the bowl. Drizzle with vinegar and scour with a toilet brush.

Moreover, not everything biodegradable is good. Take phosphates, for example. They degrade totally and quickly. Indeed, they're a potent source of nutrients for plants in water. When they get into rivers and lakes, they cause algae blooms, robbing the water of oxygen, blocking sunlight, and ultimately killing marine life. So degradability is not necessarily a panacea. As one expert put it: "Biodegradation doesn't mean anything because some products are harmful until they biodegrade."

In general, ingredients derived from petroleum oil fatty acids contain impurities that break down much more slowly. Surfactants made from natural vegetable oil fatty acids generally break down more quickly and completely. (Surfactants perform the main cleaning function in most detergents.)

But not all petrochemical-based ingredients are bad, and not all plant-based ones are good. Another expert put it this way: "Something that comes from a plant can be just as toxic as something that comes from a fossil fuel. Fossil fuels are just plants that have been sitting around for a long time."

Phosphates. The presence of phosphates in cleaners—primarily laundry and dishwashing detergents—softens water and increases the ability of surfactants to do their job. According to the Soap and Detergent Association, detergents with phosphates yield better performance per dollar. But as stated earlier, phosphates increase plant growth in water, killing water life. For that reason, they have been banned in several states.

When phosphate-free detergents were first introduced in the 1970s, they were not embraced by consumers, due to their poor performance. Consumers used twice as much detergent or washed loads twice, then switched back to brands containing phosphates. Today's versions have been vastly improved. When *Consumer Reports* tested eight phosphate and phosphate-free brands in 1987, it found little difference between the two.

Some of the phosphate substitutes cause problems, too. Zeolite, for example—used in Ecover's detergents—is relatively inert but, being insoluble, must be removed from the sedimentation process in sewage plants. Still, zeolite is among the better alternatives to phosphates at this time.

Unfortunately, a good phosphate-free automatic dishwashing detergent has yet to be invented. As a result, we did not find any phosphate-free dishwashing detergents that could be recommended to us by experts.

Bleaches. Many detergents and cleaners use chlorine (actually sodium hypochlorite) as a bleach to disinfect and remove some stains. While highly effective, when chlorine breaks down, it can react with organic and other compounds to create various

Miracle Ingredient?

If one were to come up with a list of the world's least harmful and most useful household ingredients, it would be difficult to overlook baking soda. This simple substance, tucked away in nearly everyone's pantry, may be just short of a miracle ingredient.

Baking soda is derived from a naturally occurring mineral left behind after the evaporation of an inland lake in Wyoming 50 million years ago. The mineral is converted and purified into sodium bicarbonate—baking soda. The components of baking soda—bicarbonate and sodium ions—are present in significant concentrations in the human body. Bicarbonate helps maintain proper acid/base balance in the blood, stomach, and saliva. Sodium plays a key role in the body's functioning.

Sodium bicarbonate has been sold in the United States since the 1840s and is listed by the Food and Drug Administration as a "generally recognized as safe" food substance.

Because of its unique properties and benign environmental impact, baking soda provides effective, economical, and ecological alternatives to many household cleaning chores. A few examples:

❑ **Scratchless scouring:** Sprinkle it on a damp sponge and gently scour the kitchen sink or bathroom tile, then rinse. Baking soda will even clean fiberglass sinks.
❑ **Dishwashing booster:** Add two tablespoons to the dishpan along with the usual amount of detergent. Baking soda cuts through grease and helps clean dishes faster than detergent alone.
❑ **Burnt-on food:** Sprinkle as much as needed over burnt-on spots on pots and pans. Add enough hot water to moisten the pan, then soak ten minutes (longer for stubborn spots).

substances, some of which are known to be highly toxic or even carcinogenic.

A growing number of products—including many familiar brands sold in supermarkets—have dropped chlorine bleach, usually in favor of peroxide-based bleaches. But one of the common peroxide bleaches, perborate, produces boron when it breaks down, which can be harmful to plants in some concentrations. Moreover, the process used to produce these bleaches—by reacting either borax or soda with hydrogen peroxide—requires considerable amounts of energy and raw materials.

So, "chlorine-free" does not necessarily mean a cleaning product is environmentally safe, but it is probably environmentally *better*.

Other Ingredients. Phosphates and chlorine aren't the only troublesome ingredients. Strong chemicals like ammonia, benzene, and the carcinogen NTA (often used to replace phosphates) must be filtered out of water in treatment plants. But most waste treatment plants are only about 90 percent effective. The result: Small amounts of these chemicals end up in our rivers, lakes, oceans—and in our drinking water.

Packaging. While almost all powdered detergents come in recycled cardboard boxes, liquid detergents—increasingly the preferred choice—usually come in hard-plastic bottles made of polyethylene terepthalate (PET), high-density polyethylene (HDPE), or blends of several types of plastic. While PET and HDPE are beginning to be recycled in small quantities, neither is considered an environmentally preferable material. Plastic blends, of course, are totally unrecyclable. The good news is, two big companies have begun making its jugs out of recycled material. Procter & Gamble uses at least 25 percent recycled HDPE plastic in jugs of liquid Tide, Cheer, Era, Dash, and Downy; it also uses recycled PET plastic in Spic 'n' Span Pine liquid cleaner bottles. Lever Brothers is using between 25 and 35 percent recycled resins in bottles of Wisk and All liquid detergents and Snuggle liquid fabric softener.

Many of the laundry detergents listed on pages 130-131 are packaged in plastic, though some are concentrates, which reduces packaging somewhat. Of the spray products, few products offer refills that allow you to reuse packaging. One company, EarthRite, boasts its spray bottles are made of "easily recyclable HDPE," an exaggeration at best.

Many cleaning products are packaged in aerosol cans. This, as we said previously, is an undesirable type of packaging. (See "The Problems of Packaging," page 39, for more on aerosols.) Non-aerosol plastic spray bottles have at least one advantage: they are handy for use with homemade cleaners. (See "Making Your Own," page 123.)

Cruelty-Free. Many ingredients used in household cleaners are tested on animals. Not all of this testing is necessary, and test animals are subjected to needless cruelty. A growing number of companies offer cruelty-free products—those whose ingredi-

The Wonders of Washing Soda

Its name sounds old fashioned, and indeed it has been on the market since the turn of the century (it is available under the Arm & Hammer brand). Unknown and unheralded for years, it is making a comeback. Hydrated sodium carbonate—washing soda—is derived from the same naturally occurring mineral from as baking soda (see "Miracle Ingredient?," page 125). As a natural source of alkalinity, it can serve as an effective, low-cost, and environmentally safe household cleaner. However, washing soda does not work well in homes with hard water. Here are a few suggestions on how to put it to work:

❑ **Laundry booster:** Add one-half cup along with the recommended amount of detergent. It is especially effective at cleaning large, heavily soiled loads.
❑ **Drain cleaner:** While running hot water down the drain, add a solution of one-half cup washing soda in a gallon of warm water. Then flush again with hot water. Don't try this if the drain is completely blocked, however.
❑ **Oven cleaner:** Wearing rubber gloves, sprinkle washing soda on a damp sponge and scrub surfaces to remove grease. Then rinse and dry. Do not use washing soda on aluminum surfaces.

ents use testing procedures that don't involve animals. (For more on animal cruelty, see "Personal Care Products.")

What to Look For

❑ Seek out biodegradable and nontoxic products, but don't accept these claims at face value. Look for independent verification of product claims from Green Cross or Green Seal. Or ask companies to substantiate their claims—in plain English.
❑ Buy products that don't contain phosphates and chlorine. But don't assume that a phosphate- or chlorine-free product is completely safe for the environment.
❑ Look for products that contain no artificial dyes, scents, or optical brighteners. All can be harmful to the environment.
❑ If you must use bleach, look for products that combine detergent and bleach into one product, thereby minimizing packaging.
❑ Choose non-aerosol versions of products, such as pump sprays and liquids.

❑ Buy concentrates whenever possible. Ask manufacturers to produce refillable versions that allow you to refill a spray bottle by adding water to a packaged concentrate. And ask supermarket managers to stock them.
❑ Avoid products tested on animals. Most are made of ingredients that have been around for some time. Animal testing causes great suffering to the animals involved.
❑ Look for products packaged in recycled boxes or in jugs made out of a type of plastic you can recycle in your community. At the very least, look for jugs made from recycled plastic.
❑ Buy the largest size of a product you can find to minimize packaging. Particularly avoid products with premeasured packages, one per wash load. They are extremely wasteful.
❑ Choose products from companies that have good environmental records.

Other Things to Do

❑ Use as little cleaner as possible. Whether a laundry detergent, a scouring compound, or an all-purpose cleaner, read the directions and use the minimal amount needed to do the job. Many labels overstate the amount of cleaner required; sometimes you can get good results using as little as half the suggested amount. Don't be afraid to experiment.
❑ If you have questions about specific ingredients, contact the Household Hazardous Waste Project or a local Poison Control Center (often listed in the front of your phone book). Most centers have data about chemicals' potential health hazards.
❑ Press for legislation to disclose toxins in household products, and to establish standards for environmental labeling claims.

The Bottom Line

Confused? You're not alone. Nearly every expert we talked to readily acknowledged the confounding state of affairs and the lack of easy solutions. Few products offer full disclosure of their ingredients on packages—they are trade secrets—and instead reveal only the most marketable facts: "phosphate-free," "contains no chlorine," etc. We suspect there's a lot that *isn't* being said.

Making things worse, so many federal agencies are involved—the Food and Drug Administration, the Federal Trade Commission, the Consumer Products Safety Commission, OSHA—that no one agency has a handle on the situation. Fortunately, there are a number of products that meet at least some of the above criteria. Over time, as definitions of biodegradability and other terms become standardized, it will be easier to find green brands. Certification programs, like Green Cross and Green Seal, will help.

There are some innovations in packaging. Downy fabric softener makes a refill for its 64-ounce jugs. Rather than discard the empty jug and buy a new one, you need only purchase a 22-ounce milk carton-like container. After pouring the concentrate in the jug, you add several cartonsful of water—just like making frozen orange juice—to make a new 64-ounce jug of Downy. Expect to see other such concentrates on the market soon.

In general, though, packaging is remarkably the same. Laundry detergents come either in recycled paperboard boxes or in plastic jugs, a few made from recycled HDPE plastic. Most dusting cleaners come in aerosol cans, although pump sprays are available for most. Most glass cleaners, bathroom cleaners, and all-purpose cleaners come in polyvinyl chloride (PVC) plastic containers, a type of plastic that is rarely recycled; some of these also come in aerosol form.

Here are a few things you should keep in mind about the detergent and cleaner ratings that begin on the following page:

❑ We did not list laundry detergents as "Best Choices" if they merely were chlorine- or phosphate-free, or if they were packaged in recycled cardboard boxes. There are several such brands to choose from on supermarket shelves. But if a brand had at least one other redeeming factor—for example, if its manufacturer rated highly, or it was made without animal testing—we included it.

❑ Similarly, we did not list products as "Worst Choices" simply because they were packaged poorly. Again, many products come in plastic jugs, which we consider undesirable, even if they contain some recycled material; cardboard packaging is still preferable. The only exception were spray starches, which we listed as "Worst Choices" because of their aerosol packaging.

❑ Many of the smaller, alternative cleaner manufacturers were not among those rated for their companies' environmental records, although several likely would have rated highly. We hope to include these companies' ratings in future editions.

Laundry Goods

Best Choices

Any concentrated detergent.
Any detergent in a recycled cardboard box.
Ajax (box) (Colgate-Palmolive)
All Concentrated Detergent (Lever Brothers)
Allen's Naturally products (Allens Naturally)
Arm & Hammer Heavy Duty Det. (box) (Church & Dwight)
Arm & Hammer Super Washing Soda (Church & Dwight)
Axion Bleach Alternative (box) (Colgate-Palmolive)
Bold Detergent & Softener (Procter & Gamble Co.)
Clorox Bleach (box) (Clorox Co.)
Downy Refill (Procter & Gamble Co.)
Dynamo (Colgate-Palmolive)
Earth Wise Laundry Stain Remover (Earth Wise Inc.)
Eco-Force Liquid Laundry Detergent (Ecomer Inc.)
Ecover products (Mercantile Food Co.)
Fab (box) (Colgate-Palmolive)
Fresh Start (Colgate-Palmolive)
King Liquid Laundry Starch (Wonder Chemical Corp.)
Life Tree products (Life Tree Products)

Best Choices (Cont.)

Shout (S.C. Johnson & Son Inc.)
Ultra Cheer (Procter & Gamble Co.)
Ultra Oxydol (Procter & Gamble Co.)
Ultra Tide (Procter & Gamble Co.)
Winter White Laundry Detergent (Mountain Fresh)

Worst Choices

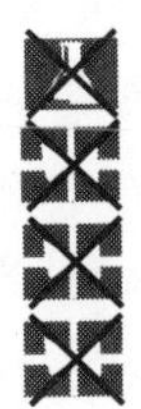

Any product containing phosphates or chlorine bleach.
Any product in a nonrecyclable plastic jug.
Spray 'n' Starch (DowBrands Co.)
Spray 'n' Wash (DowBrands Co.)

Household Cleaners

Best Choices

Ajax (Colgate-Palmolive)
Allen's Naturally products (Allen's Naturally)
Arm & Hammer Baking Soda (Church & Dwight)
Arm & Hammer Carpet/Room Deodorizer (Church & Dwight)
Bon-Ami (Faultless Starch/Bon Ami Co.)

[continued]

Best Choices (Cont.)

Brite (S.C. Johnson & Son Inc.)
Citra-Solv Natural Citrus Solvent (Chempopint Products Co.)
Dermassage (Colgate-Palmolive)
EarthRite cleaners (EarthRite)
Earth Wise products (Earth Wise Inc.)
Ecover products (Mercantile Food Co.)
Fels Naptha (Dial Corp.)
409 (Clorox Co.)
Future (S.C. Johnson & Son Inc.)
Handi-Wipes (Colgate-Palmolive)
Kleen All-Purpose Cleaner (Mountain Fresh)
Life Tree products (Life Tree Products)
Livos PlantChemistry products
Murphy's Oil Soap (Murphy-Phoenix Co.)
Natural Chemistry products (Natural Chemistry Inc.)
Palmolive (Colgate-Palmolive)
Pledge (S.C. Johnson & Son Inc.)
S.C. Johnson Paste Wax (S. C. Johnson & Son Inc.)
Simple Green Concentrated Cleanser (Sunshine Makers)
Spic 'n' Span Concentrated Powder Formula (Procter & Gamble Co.)
Tile-X Instant Mildew Stain Remover (Clorox Co.)
Winter White Glass Mate (Mountain Fresh)

Worst Choices

Any aerosol product.
Any product containing phosphate.
Dow Bathroom Cleanser with Scrubbing Bubbles (DowBrands Inc.)
Easy-Off (Boyle-Midway Household Products)
Glass Mates Glass Cleaning Wipes (Lehn & Fink Prod. Group)
Lestoil Deodorizing Rug Shampoo (Noxell Corp.)
Lysol Basin Tub & Tile Cleaner (Lehn & Fink Products Grop)
One-Step No-Buff Wax (S.C. Johnson & Son Inc.)
Resolve Carpet Cleaner (Lehn & Fink Products Group)
Spiffits Cleaner (DowBrands Inc.)
Spray 'n' Vac Rug Cleaner (aerosol) (Glamorene)
Spruce-Ups Moist Household Wipes (American Cyanamid Co.)
Woolite Rug Cleaner (aerosol) (Boyle-Midway Household Products)

Kitchen Wraps and Bags

In many respects, kitchen wraps and bags are some of the most environmentally—and economically—important purchases we make. By effectively protecting leftovers and other stored foods, materials such as aluminum foil, plastic wrap, and wax paper save food—and money—from being thrown away. While only foil is both recyclable and contains recycled material, other materials, used sparingly, are good buys.

What's the Problem?

It's when these materials are destined for disposability that they create both opportunities and challenges for Green Consumers.

One of the big problems—and perhaps the first major Green Consumer casualty—has to do with degradable trash bags. In 1989, Mobil Chemical Co. (which makes Hefty brand trash bags) and First Brands (Glad brand bags) introduced "biodegradable" trash bags. Other companies soon introduced similar products. (Some were known as "photodegradable," others simply "degradable.") It was such a seductive notion: By adding just a few teaspoons of corn starch to the recipe for plastic bags, we could make them virtually melt away in the noonday sun.

Alas, it was not to be. While degradable under laboratory conditions, they did not simply "melt away" in a landfill, where there is little oxygen or sunlight to facilitate the degrading process. Landfills, as we've learned, tend to mummify trash—including trash bags—better than they dispose of it.

It wasn't as if these degradable trash bag makers had simply miscalculated. Indeed, some of them admitted that they didn't really believe the degradability claims, but were using it as a marketing tool because that's what people were asking for. The Federal Trade Commission and a task force of state attorneys general took a harsh view of this behavior, and levied fines against several companies.

What to Look For

❑ When buying trash bags, ignore claims of degradability. Look for bags that contain at least some recycled plastic.

❑ When buying kitchen wraps, buy the largest size package you can to minimize packaging.

❑ Buy sandwich bags made of cellulose, which is made from wood pulp, instead of plastic, which is made from petroleum.

Other Things to Do

❑ When covering foods before microwave cooking, use wax paper instead of plastic wrap. Wax paper allows steam to escape, reducing the chance you'll burn yourself. And when plastic wrap comes into contact with food while microwave cooking, there is a chance some of the chemicals in the wrap may leach into the food.

❑ Try to minimize your use of kitchen wraps by keeping stored foods in resealable and reusable plastic containers.
❑ Recycle as much as possible, so you'll use fewer trash bags.

The Bottom Line

There are few differences among kitchen wraps and bags, with three exceptions: Sandwich bags made from cellulose instead of plastic, trash bags made using at least some recycled plastic, and products from environmentally responsible companies. Unfortunately, we could not find any national brands of cellulose bags; they can be found in a few health food stores and from mail-order companies. There are only two companies, Earth Wise Inc. and Webster Industries, that currently make trash bags with recycled plastic. And there were no companies that rated highly for environmental performance.

Best Choices

Earth Wise Trash Bags (Earth Wise Inc.) 
Good Sense trash bags (Webster Industries)
Renew Trash Bags (Webster Industries) 

Hefty trash bags (Mobil Chemical Co.) 

Paper Products

Paper is the number one ingredient in our landfills, and we generate a lot of it at home. Packaging, newspapers, junk mail—all contribute to the two or three pounds of paper trash the average American household discards every day.

But it's not just the amount of paper that is of concern. Considering how they're made, some of our most basic household paper products can be downright deadly.

What's the Problem?

One big problem has to do with the bleaching process used in making many paper products, including paper towels, toilet paper, facial tissues, dinner napkins, tampons, tea bags, coffee filters, disposable diapers, even milk cartons.

The bleaching process that makes these papers white uses chlorine gas. Making certain kinds of paper involves a five- or six-stage bleaching sequence to achieve high brightness. One byproduct of that bleaching process are dioxins, a deadly chemical that in one form was used as "Agent Orange," a defoliant weapon used in Vietnam. Dioxins are extremely toxic and cancer-causing. Even minute quantities can trigger anything from acne and achy joints to insomnia and immune system disorders. It is important to keep in mind that *the smallest detectable amount of dioxins have been known to cause cancer in laboratory animals.*

The dioxins created in pulp making don't just end up in the paper products. In fact, scientists first made the link between dioxin and paper after discovering unexpectedly high levels of the chemicals in fish downstream of pulp and paper mills. The high dioxin levels in the fish made them unsafe for human consumption. Moreover, when dioxin-contaminated water is used to irrigate crops, they, too, become contaminated. The Environmental Protection Agency reported in 1988 that, when grown in contaminated soils, root crops—such as carrots, potatoes, and onions—can develop dioxin levels that equal or exceed those found in the soil itself.

The paper products themselves present problems, too. An official at the Food and Drug Administration told *Science News*

Paper Products to Avoid

Another way to reduce the amount of paper in our trash is to avoid buying paper products that cannot be reused or easily recycled. Here are some examples:

❑ **Paper Cups and Plates** are coated with wax or plastic to make them impervious to water. They are neither biodegradable nor recyclable. They cause the same disposal problems as Styrofoam (polystyrene) cups and plates. A better solution is reusable china, enamel, or sturdy reuseable plastic.

❑ **Window Envelopes** contain plastic and synthetic glue, which can upset the recycling process. Don't use these unless you are sure the window is made of cellulose or contains corn starch.

❑ **Coffee Filters** create an endless stream of paper waste. If your coffee maker supports the use of a gold filter, use this nondisposable type. Reusable cotton filters are also available. Another alternative is to use a plunger-type coffee maker.

in 1989 that the two major sources of dioxins were milk cartons and coffee filters. The official's self-described "very rough estimates" were that young children getting all their milk from contaminated cartons might double their dioxin intake. Heavy coffee drinkers consuming most of their brew from pots with bleached-paper filters might increase their daily dioxin intake 5 to 10 percent above the average U.S. level.

That's not all. When these paper products are disposed of in a landfill or are incinerated, the dioxins released in the air can be inhaled by animals and humans or ingested through contamination of food crops.

Interestingly, paper products made from recycled pulp don't have dioxin problems, either in the products themselves or in the wastewater created downstream. The reason has to do with the fact that most of the wood juices that mix with chlorine during paper making have already been cooked out by the time paper is recycled. And the levels of chlorine used to bleach recycled pulp are lower than for virgin pulp; some recycled paper companies use alternative bleaching systems that use hydrogen peroxide, among other formulas. So, buying paper products made from recycled content serves two purposes: it helps to close the recycling "loop" and minimizes dioxin released into our water supply and paper products.

What to Look For

❑ Buy paper products made of 100 percent recycled material.
❑ Choose unbleached paper products, or those bleached using non-chlorine bleaches.
❑ Avoid paper products that you cannot recycle in your community.
❑ Choose products from companies with good environmental records.

Other Things to Do

❑ Reduce your use of paper products altogether. For example, substitute rags, sponges, and cloth towels for paper towels. Substitute cloth diapers for disposable ones. (See "Baby Products" for more on disposable diapers.) Use a permanent, reusable metal coffee filter rather than disposable paper ones.

The Bottom Line

There is a growing number of recycled paper products on supermarket shelves. But read labels carefully: Make sure the label states 100 percent recycled content. While most stores carry at least one brand, many stores do not. If yours doesn't, ask the manager to carry them. Note that most of the companies listed on the next page offer an entire line of paper products—paper towels, toilet paper, facial tissue, napkins, etc.

Several companies in this category rated poorly on environmental performance, testimony to the amount of pollution created in the paper-making process.

Best Choices

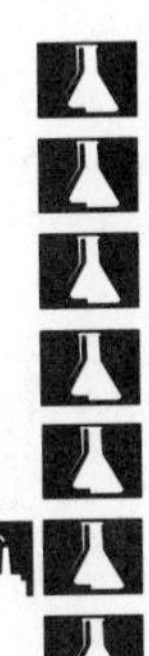

C.A.R.E. paper products (Ashdun Industries Inc.)
Enviro-Care paper products (Ashdun Industries Inc.)
Envision paper products (Fort Howard Corp.)
Green Forest paper products (Fort Howard Corp.)
Marcal paper products (Marcal Paper Mills Inc.)
Melitta Natural Brown coffee filters (Melitta N. America Inc.)
Natural Brew coffee filters (Rockline Inc.)
Today's Choice paper products (Confab Industries)

Worst Choices

Big 'n' Pretty napkins (Georgia-Pacific Corp.)
Big 'n' Soft bathroom tissue (Georgia-Pacific Corp.)
Big 'n' Thirsty paper towels (Georgia-Pacific Corp.)
Brawney paper towels (James River Corp.)
Coronet Bathroom Tissue (Georgia-Pacific Corp.)
Cottonelle Toilet Tissue (Scott Paper Co.)
Kleenex products (Kimberly-Clark Corp.)
Northern products (James River Corp.)
Scott and Scotties products (Scott Paper Co.)
Vanity Fair All-Occasion Napkins (James River Corp.)
Viva products (Scott Paper Co.)

Personal Care Products

We won't even delve into the issue of whether we truly need all those cosmetics and toiletries found on supermarket and drugstore shelves these days. Let's assume for the moment that all these products are among life's necessities. Even at that, there's a tremendous amount of waste in some of these products.

What's the Problem?

It doesn't take a private investigator to scan the personal care products shelves and see the tremendous amount of packaging surrounding these products. The personal care products industry are masters of packaging, creating ingenious shapes and sizes to make their goods both convenient and eye-catching.

Products marketed to women are particularly overpackaged. Just compare the same brand of deodorant marketed to each gender. In many cases, the women's version (the contents of which is usually no different than the men's version) features at least one extra layer of packaging. Why? It has something to do with manufacturers' assumptions about the way women choose their purchases. Do they really believe women desire extra packaging? Apparently so.

Plastic reigns supreme in the personal care section. Part of that has to do with the fact that most of these products are intended for use in the bathroom, where broken glass is a safety hazard. And there are plastic safety seals on many of these products, the result of our terror-driven society.

But this isn't just any plastic. Most plastic shampoo bottles and deodorant containers are made of "composite" materials—multiple layers of different plastics that are impossible to recycle. A few bottles are made of PET or HDPE plastic (see "Which Plastic Is It?," page 35, for more on this), which is being recycled in small but growing amounts. But most Americans do not yet have the ability to recycle these materials.

Toothpastes are another area of overpackaging—at least those that come as disposable "pumps." Pumps contain three times as much plastic as an old-fashioned toothpaste tube for the same amount of dentifrice. Of course, these things are not recyclable.

Still another packaging problem for many personal care products are aerosols. Hair sprays, deodorants, and antiperspirants are the products most frequently packaged as aerosols, but there are others—foot sprays, first-aid sprays, and insect repellants, for example. (For more about aerosols, see "The Problems of Packaging," page 39.) Fortunately, all these products are available as nonaerosols—pump sprays, roll-ons, sticks, powders, liquids, lotions, and creams. Ruling out aerosols altogether makes your purchasing decision easier from the start.

Disposables. With personal care products, it isn't just the packages that are wasteful. For years, disposability of the products themselves has been a big selling device for personal care companies. From razor blades to tampon applicators, "use it and lose it" has been the theme. And the price tag for all that convenience is finally coming to light, from overcrowded landfills to plastic-strewn beaches (plastic tampon applicators have become known derisively as "New Jersey seashells").

Advertisers point to the busy lifestyles of Americans when defending the throwaway mentality they have inspired and promoted. They tell us that they are giving us what we want: convenience. They explain that we simply don't have time to be bothered with refilling, reusing, or recycling the products we buy; we'd much rather leave these nasty details to others.

Perhaps. But what about the "nasty details" of the environment? Marketers' perceptions notwithstanding, studies indicate that many of us *do* want to be bothered with these matters, that we don't want to knowingly contribute to the earth's problems. Maybe, just maybe, the conveniences offered by our throwaway society are as temporary as the products themselves.

Contents. We are just beginning to look *inside* many of the personal care products we have purchased for years. And what we are finding is not always a pretty sight. Many of the products designed to cleanse, deodorize, and otherwise improve our bodies contain ingredients that are not always healthy to people or the planet. Some, like formaldehyde (found in shampoos, deodorants, and even toothpaste) are suspected carcinogens. Other ingredients, including ammonia, ethanol, and many of the colors and fragrances used, can be irritants or produce allergic reactions.

Chlorine bleaching is another problem, associated in this case with feminine hygiene products. While major tampon producers such as Tambrands (makers of Tampax) and Johnson & Johnson (makers of o.b. tampons) have focused on disposability of their products—both have introduced "biodegradable" and "flushable" products—none of the major companies have changed the way their products are bleached. (See "Paper Products" for more on how chlorine bleaching leads to the formation of toxic compounds such as dioxins.) There are non-chlorine bleached products on the market, although they are rarely available in supermarkets.

Cruelty-free. The idea of "cruelty-free" products is a relatively new one, and a confusing one. On the one hand, consumers want products that are safe to use, that won't cause irritations or other harmful or allergic reactions, and that won't harm small children who ingest them. On the other hand, there is a growing concern over the treatment of animals used in testing these products.

But as with so many other "green" issues, the idea of what exactly is "cruelty free" can be murky at times. And some companies have been less than forthcoming on whether and how they are using animals to test their products.

Millions of laboratory animals—mice, rats, dogs, cats, monkeys, and others—sacrifice their health, and often their lives, in the name of "science," testing everything from cleaners to cosmetics to children's toys. Before it reaches your supermarket shelves, most "new" or "improved" products go through a battery of animal tests to make sure they are safe for humans. In the process, some 14 million animals die each year. Over 50 million more are put through some kind of testing, the result of which may be the animals' painful disfigurement.

Some tests are brutal. Lab animals are routinely burned, injected with poisons, electrically shocked, and subjected to other forms of suffering. One notorious test, the Draise Eye-Irritancy Test used to measure the eye irritancy of ingredients, involves putting albino rabbits in restraining devices, then administering a few drops of the test substance into their eyes.

Finding Cruelty-Free Products

More than 300 companies currently make products that are not tested on animals, according to People for the Ethical Treatment of Animals (PETA), although only a handful of major national brands appear on the list, including Amway, Aramis, Aveda, Avon, Clinique, Estee Lauder, Max Factor, and Revlon.

Sorting through the hundreds of companies is made easier with PETA's publication, *Shopping Guide for Caring Consumers.* The publication includes lists of companies that don't test on animals, what types of products they carry, and where the products are available, among other information. To order, send $2.95 to PETA, P.O. Box 42516, Washington, DC 20015.

The Humane Society of the United States (HSUS) also offers a name-brand guide—a wallet-size "Beautiful Choice Directory." Cost is $2.50 for 25 copies, $3.50 for 50 copies. Write to HSUS, 2100 L St. NW, Washington, DC 20036.

The frequent result: the animals' eyes swell and redden until they go blind.

There is considerable controversy over the validity of the tests. Animal-rights activists claim that most such tests are needless, and that there are testing methods available that don't involve animals. In 1990, the National Academy of Sciences began examining the usefulness and effectiveness of animal testing. A paper published in the prestigious journal *Science* questioned whether the tests adequately provide a measure of ingredients' effects on humans. As one researcher put it, "We spend hundreds of millions a year chasing after these little traces of things that may be no risk at all." Many product manufacturers counter that animal testing is the only truly reliable assurance of a product's safety.

Misleading Terms. The term "cruelty free" can mean different things to different people. According to experts, the term is used rather loosely. For some products, it can mean that the ingredients were not tested on animals. For others, it means the product contains no animal ingredients. For still others, it means only that the company has declared a moratorium on animal testing, or didn't test on that particular product, but still tests other products on animals, or at least hasn't ruled out resuming testing in the future.

Identifying cruelty-free products has become easier with the help of logos from two organizations. People for the Ethical Treatment of Animals (PETA) has two logos available to products that are cruelty free, one signifying that a product wasn't tested on animals, the other stating that the product also contains no animal ingredients. A third logo comes from the Humane Society of the United States (HSUS), part of the group's "Beautiful Choice" campaign. The HSUS logo attests only to a lack of animal testing, not to a lack of animal ingredients.

The Beautiful Choice™

With awareness about testing on the rise, many companies have tried to assure consumers that their products are kinder and gentler to animals. But not all are being completely honest.

Several companies have announced moratoriums on testing, giving the impression they are cruelty free, but without necessarily renouncing further testing. Mary Kay Cosmetics, for example, announced such a moratorium, but Richard C. Bartlett, the company president, in a letter to Chemical and Engineering News, later pointed out that "I specifically did not say that [we] would stop all animal testing forever." According to PETA, such companies as Merle Norman, Noxell, Faberge, Armour-Dial, and BeautiControl have "capitalized on positive press by announcing a moratorium on animal tests but have refused PETA requests that they ban these cruel tests permanently."

What to Look For

❑ Choose products that are minimally packaged, and whose packaging can be refilled or recycled in your community.
❑ Buy products made with organic ingredients, which minimize the use of polluting pesticides and fertilizers.
❑ Ask for non-chlorine bleached feminine hygiene products. Buy tampons that require no applicator; if an applicator is needed, opt for cardboard over plastic.

Who Tests, Who Doesn't

Here is a select list of personal-care products and companies that do not conduct animal testing, and those that do. For a complete list, write for the PETA guide listed on page 143.

Companies that Do Not Test on Animals

Almay
Aramis
Aveda
Avon
Bare Essentials
Bill Blass
Charles of the Ritz
Christian Dior
Clinique
Crabtree & Evelyn
Estée Lauder
Flex
Jean Naté
Kiss My Face
Max Factor
Revlon
Tom's of Maine

Companies that *Do* Test on Animals

Alberto Culver
Bristol-Myers
Chanel
Clairol
Colgate-Palmolive
Gillette
Helena Rubenstein
Helene Curtis
Jergens
Jovan
Lancome
L'Oreal
Maybelline
Mennen
Neutrogena
Sea & Ski
Vidal Sassoon
Wella

❑ Choose nonaerosol products, such as pump sprays, sticks, and roll-ons.
❑ Avoid single-use and disposable products. They are wasteful.
❑ Buy the largest size of a product you can find to further minimize packaging.
❑ Avoid products tested on animals. Most are made of ingredients that have been around for some time.
❑ Choose products from companies that have good environmental records and that don't test on animals.

The Bottom Line

The personal-care aisle of most supermarkets offers mostly major brand-name items, few of which are responsibly packaged or have contents that meet the criteria above. Products marketed specifically to women—the majority of them are—seem to have the most excessive packaging. We found several instances where the women's version of a product, such as a deodorant, had at least one more layer than the man's version of the same product.

To find the greenest personal care products, you'll probably have to check with health food stores, specialty stores such as The Body Shop, or mail-order companies (see "Shopping by Mail," page 182). Moreover, there were few mainstream companies with top environmental records whose products seem to be readily available in supermarkets. (One company with a good record, Gillette Co., conducts animal testing and has been a principal target of animal-rights groups.) Indeed, only one company, Tom's of Maine, rates highly for both company, packaging, and contents.

Best Choices

Agree Shampoos & Conditioners (Carol Hansen)
Aveda products (Aveda Corporation)
Carefree Panty Shields (Johnson & Johnson)
Halsa Shampoos & Conditioners (Carol Hansen)
J.R. Liggett's Old Fashioned Bar Shampoo (J.R. Liggett Ltd.)
o.b. Tampons (without applicator) (Johnson & Johnson)
Selson Blue Dandruff Shampoo (Ross Laboratories)
Stayfree Maxi Pads (Johnson & Johnson)
Sure and Natural Panty Liners (Johnson & Johnson)
Today's Choice Maxi Pads (Confab Sales & Marketing)
Today's Choice Panty Liners (Confab Sales & Marketing)
Tom's of Maine products (Tom's of Maine)

Worst Choices

Any product tested on animals.
Kotex Security Tampons (Kimberly-Clark)
Light Days by Kotex (Kimberly-Clark)
o.b. Tampons (with applicator) (Johnson & Johnson)
New Freedom Any Day Panty Liners (Kimberly-Clark)
Playtex products (Playtex Inc.)
Tampax Tampons (HyGeia Sciences Inc.)

Pet Foods

What are we feeding our pets? No doubt, it's a nutritionally balanced meal; the science of pet nutrition has improved considerably in recent years.

But what's in those products, environmentally speaking?

It's a question that's not easy to answer. For example, there's little information available about the source of the meat used to feed our fussy cats and dogs. Is it part of the lower-quality beef that is imported from rain forest regions—and for which forested areas were slashed and burned to make grazing areas? Companies won't tell; these are trade secrets.

Packaging of pet foods is a relatively straightforward issue. Canned pet foods are generally responsibly packaged in recyclable steel cans. Most dry foods come in multilayer bags to maintain freshness; there is probably little way around this. Probably the worst packaging are the individual servings intended for cats, as well as the so-called premium moist dog foods. Like individual servings of food for humans, these are usually a waste of packaging and an added expense, and there are almost always reusable containers available that can serve the dual needs of freshness and convenience. As always, the greenest choice is to buy the largest size package available.

The biggest differences we found were among the environmental records of the major pet-food companies. Several leading companies, including Quaker Oats and Heinz, are among the top-rated firms. It is here that Green Consumers can cast their green vote with every purchase.

Best Choices

Amore (Heinz Pet Products)
Cat Liner Pads (Soft Soap Enterprises Inc.)
Control Cat Litter (Clorox Co.)
Fresh Steps Cat Litter (Clorox Co.)
Gaines products (Quaker Oats Co.)
Gravy Train (Quaker Oats Co.)
Ken-L Ration (Quaker Oats Co.)
Kibbles 'n' Bits (Quaker Oats Co.)
Kozy Kitten (Heinz Pet Products)
Meaty Bone (Heinz Pet Products)
Moist Meals (Quaker Oats Co.)
9 Lives (Heinz Pet Products)
Puss 'n' Boots (Quaker Oats Co.)
Recipe dog food (Heinz Pet Products)
Vets dog food (Heinz Pet Products)

Worst Choices

Bonkers (Nabisco Brands Inc.)
Gaines Burgers (Quaker Oats Co.)
Milk Bones (Nabisco Brands Inc.)

Part Three

T·H·E G·R·E·E·N S·U·P·E·R·M·A·R·K·E·T

THE GREEN SUPERMARKET

The supermarket industry did not want us to write about them in this book. In fact, an official at the Food Marketing Institute (FMI), the principal trade association of supermarkets in the United States, personally wrote to the major supermarket chains and urged them not to provide us with information about their environmental policies. After being contacted by FMI, a few chains even tried to "withdraw" the information they had provided us. (See, "Are These Guys Hiding Something?," page 164, for names and addresses of major chains that would not reveal their environmental policies and practices.)

Frankly, we're not sure what they are afraid of. Many stores are doing some pretty impressive things to help their customers help the environment. Several have developed extensive environmental policies that spell out their commitment to being responsible companies. Others have launched ambitious information campaigns to help turn customers into Green Consumers.

So, why is the supermarket industry afraid of telling us what they're up to? Perhaps because while some stores are taking the lead, others aren't doing much at all.

The New "Green" Grocer

It used to be that "green grocer" referred to someone who sold produce plucked right out of the ground—or at least right off a farmer's truck. The term has come to mean something entirely different. A "green" grocer is one who subscribes to the notion that environmental responsibility begins at home, and much of what we bring home comes from the supermarket. So, today's green grocer takes the Earth into account in a variety of ways, from the way stores operate to the kinds of products on the shelves and the way they are labeled.

The changes may not be readily apparent, but they are sprouting in store after store. As with so many other industries, grocers are recognizing that there's plenty of "gold" in all that green—an opportunity to do well by doing good. Many supermarket chains are finding that being environmentally responsible pays off in at least two ways. First and foremost, it's good marketing, a way to lure the growing numbers of Green Consumers. One grocer, J.B. Pratt, has turned his 10-store Oklahoma chain, Pratt's, into an "enviromarket." Full-page newspaper ads—in green ink, of course—promote greener products and even discourage customers from buying so-called "biodegradable plastics."

But many "green" operating procedures and policies are also good for the bottom line, reducing waste, increasing efficiency, motivating employees, cutting costs, and increasing profits. From recycling wastes to increasing the efficiency of their refrigeration systems, there are many opportunities to turn responsible practices into good profitable business.

We think it's the least stores can do. And when you think about it, supermarkets have a responsibility to help us become greener consumers. At least half of the trash we generate at home is purchased there, producing many of the pollutants we wash down the drain. After decades of contributing to environmental problems, it's time for grocers to start contributing to solutions.

What Makes a Supermarket Green?

In consultation with industry experts and supermarket executives, we've identified twelve factors that represent the state of the art of supermarket environmental practices:

1. Company Environmental Policy. The supermarket chain should have a written environmental policy stating its intentions and goals on environmental matters.

2. Consumer Information. Stores should offer environmental shopping information to consumers, helping them to make "green" choices.

3. Shelf Labeling. Stores should provide reliable shelf labeling to help consumers find products that are responsibly packaged or that have other environmentally desirable attributes.

4. Shopping Bags. Stores should give customers a choice between paper and plastic shopping bags, and should take both types of bags back from customers for reuse or recycling. Also, stores should make reusable canvas bags available to consumers at a reasonable price.

5. Recycling Drop-off. In communities that do not have curbside pickup or other centralized recycling facilities, stores should offer consumers places to deposit cans, bottles, newspapers, and other materials for recycling. (In communities that do have curbside recycling programs, such supermarket efforts are considered to be counterproductive to city and county collection efforts, and are therefore not desirable.)

6. Meat Departments. Supermarket meat departments or butcher shops should offer paper wrapping as an alternative to plastic packages. In addition, meat scraps and bones, instead of being discarded, should be used by the stores for rendering, soup stocks, sauces, or composting.

7. Bulk Purchases. Supermarkets should encourage the purchasing of bulk grains and other foods, as well as case-lot quantities of packaged goods.

8. Organic Produce. Supermarkets should offer organic, transitional, or pesticide-free produce.

9. CFC Recovery. Stores should use systems that recover and recycle ozone-damaging chlorofluorocarbons (CFCs) in their refrigeration, freezer, and air conditioning systems.

10. Supplier Conditions. Supermarkets should encourage suppliers to reduce or eliminate unnecessary packaging, or to use packaging that can be recycled or reused.

11. Internal Recycling. Stores should recycle their own cardboard, glass, aluminum, plastic, and other materials. Food scraps should be composted.

12. Energy Conservation. Stores should actively conserve water, energy, petroleum, and other resources within their facilities and vehicles.

Granted, these twelve items are no small order. And just as there are few if any perfectly green products, there are few perfectly green supermarkets. Still, we found many good examples of company policies and practices.

It should also be pointed out that consumer behavior does not always support some of these green supermarket policies. For example, for all the concern over pesticides in foods, many supermarkets reported that they have stopped selling organic fruits and vegetables because of consumer lack of interest. (Organic farmers and distributors respond that the supermarkets didn't actively promote organics, or hid them off in some "organic ghetto" in a corner of the store. In any case, this "green" policy has not succeeded in some supermarkets.)

In our descriptions of the leading supermarket chains' policies on the above factors, it is important to understand that some of these questions do not warrant simple yes-or-no answers. Consumer-information or shelf-labeling programs, for example, vary widely in quality among stores, and the mere existence of such programs does not mean that they are good or effective. We have tried to summarize programs as accurately they existed in early 1991.

The Major Chains

Following are summaries of the environmental policies and practices at most of the nation's major supermarket chains, as reported by each company. If you don't see your supermarket in these tables, check the address list beginning on page 162; some companies operate stores under several names. (See "Are These Guys Hiding Something?," page 164, for names of supermarkets that refused to provide information.) Because some of the greenest supermarket chains we found were not among the biggest, we have also included descriptions of some smaller but innovative companies, as well as a few wholesalers that supply exclusively to several chains. Note that many of the organizations we talked with were planning various activities, so this information is subject to change.

See page 164 for information on how to rate supermarkets not on this list.

	A&P	Albertson's	Bi-Lo
Company Policy	No formal policy	No formal policy	Published a comprehensive 12-point agenda
Consumer Information	Developing brochures		Offers own brand of recycled paper products, detergents, and other products
Shelf Labeling	Labels appear on 250 products that meet environmental standards		
Shopping Bags	Offers paper and plastic; some bags contain recycled content. Most stores sell canvas bags	Offers paper and plastic; sells cloth bags. All bags made from partly recycled materials. 5-cent rebate given for returned bags	Offers paper and plastic; some paper bags made of recycled content. 5-cent rebate on returned paper bags
Recycling Drop-off	Aluminum drop-off where curbside pickup not available		Paper and plastic bags
Meat Department	Choice of paper or plastic wrapping in most stores	Choice of paper or plastic wrap. Introducing alternative packaging for meats	
Bulk Purchases	Available in most stores		
Organic Produce	Available where customer demand warrants	Discontimued due to lack of customer interest	
CFC Recovery	Recovery systems in all stores		
Supplier Conditions	Suppliers asked to back claims made by products they carry	Sent 13,000 letters to suppliers urging involvement	Encourages source reduction and recycling
Internal Recycling	Computer paper, white paper, aluminum, glass, plastic	Newspapers, aluminum, glass, computer paper, cardboard, plastic film	Cardboard, motor oil, office paper
Energy Conservation	Programs to conserve refrigeration, heat, and air conditioning in all stores		Refrigeration waste heat provides store heating

	Big Bear	Bread & Circus	Bruno's
Company Policy	No formal policy, but extensive enviromental advertising	No formal policy	Formal written policy
Consumer Information		Offers pamphlets on organic farming and select products; also educates in advertising	Monthly brochure highlights environmental topics
Shelf Labeling	Shelf labels promote environmentally safe ideas	Labels emphasize recycled packaging and other environmental benefits	
Shopping Bags	Promotes paper over plastic, but offers recycled plastic bags. Sells canvas; donates profits. 4-cent rebate for used bags	Offers paper and plastic; 3-cent rebate on returned paper bags	Offers paper and plastic; plastic bags made of part recycled material. Sells canvas bags
Recycling Drop-off	Aluminum, glass, and plastic	Plastic bags, aseptic packaging	
Meat Department	Wraps special orders in paper; plastic wrap includes recycled material	Wraps meat in recycled paper and uses paper trays	Uses plastic and paper wrapping
Bulk Purchases	Offered at a discount	Large bulk section	
Organic Produce	Wide variety available	Wide variety available	Available in stores where consumer interest sufficient
CFC Recovery	Efficient, up-to-date equipment		Uses recovery and recycling systems
Supplier Conditions	Asked all suppliers for environmental policy and list of safe products	Notifies suppliers when packaging in excess; drops products that don't comply	Encourages elimination and reduction of unnecessary packaging
Internal Recycling	Newspaper, glass, plastic, computer paper, cardboard, aluminum, wooden pallets, and vehicle oil and coolants	Bottles, cans, all paper	Corrugated cardboard, white office paper
Energy Conservation	Encourages stores to conserve through cash prizes	Operates under maximum energy efficiency	

	Dominick's	First National	Food Lion
Company Policy	No formal policy	No formal policy	No formal policy, but company "trying to respond to customer requests"
Consumer Information		Has education program built around "3 R's"; provides recycling information hotline	
Shelf Labeling		Signs offer environmental tips but do not rate products	
Shopping Bags	Offers paper and plastic and reusable shopping bags	Offers paper and plastic; sells low-cost canvas and mesh bags	Offers paper and plastic; paper bag made from 100% recycled material
Recycling Drop-off	Plastic grocery bags; works with communities on yard waste disposal project	Paper and plastic bags	Paper and plastic bags
Meat Department			
Bulk Purchases			
Organic Produce			
CFC Recovery			Refrigerant recovery systems in all stores
Supplier Conditions			
Internal Recycling	Aluminum cans, waste paper	Corrugated cardboard	Cardboard, all office paper; reuses motor oil; recaps tires on trailers
Energy Conservation			

	Giant Food	Giant Eagle	Grand Union
Company Policy	Has environmental task force and published policy statement	Formulating	Plans to release formal policy statement
Consumer Information	Provides information on a variety of topics	Brochures on green topics; hosted public TV show on environmental shopping	
Shelf Labeling			
Shopping Bags	Offers paper and plastic; sells mesh bags at cost	Offers paper and plastic; sells cloth bags. 5-cent rebate on all returned bags	Offers paper and plastic; sells cloth bags. 5-cent rebate for returned plastic bags
Recycling Drop-off	Some stores accept newspapers, plastic bags, aluminum	Plastic bags	
Meat Department		Offers paper and plastic wrapping	
Bulk Purchases			Discounts available for bulk purchases
Organic Produce		Offered in largest stores	Stocked periodically when supply and demand are adequate
CFC Recovery		Has leak-detection system	
Supplier Conditions	Encouraging suppliers to reduce packaging where possible	Encourages source reduction	Redesigned private-label products to reduce packaging. Encourages national brands to reduce packaging
Internal Recycling	Cardboard, white paper, computer paper, aluminum cans, plastic film wrap; trailers use recapped tires	Cardboard, computer paper, aluminum	Aluminum, newspapers, computer paper; plans for polystyrene
Energy Conservation	Natural gas being tested in fleet vehicles. Benches and parking lot bumpers made from recycled plastic	Proposing hand dryers and faucet aerators	Heat reclamation, energy management program, extra insulation, and energy-efficient lighting in all stores

	Kroger	Fred Meyer	Pathmark
Company Policy	No company-wide policy; each store sets own policies	Has extensive written policy for all company personnel	In progress
Consumer Information	Programs in some regions. Example: Atlanta stores have public education booths	Brochures in all stores; recycling information on grocery bags	Used bag stuffers as Earth Day tie-in
Shelf Labeling		Uses Green Cross labeling program; features own "earth-friendly" brands	
Shopping Bags	Most stores offer paper or plastic and sell canvas bags. All give rebates for returned paper bags	Uses mostly paper bags with recycled content, though some departments use plastic; sells canvas bags	Offers paper and plastic; sells mesh bags. Paper bags made from recycled content
Recycling Drop-off	Plastic bags, aluminum, glass, newspapers		Plastic bags in Babylon, Long Island, store only
Meat Department	Deli departments recycle grease; meat departments reprocess meat trimmings	Uses both paper and plastic wrapping	
Bulk Purchases		Encouraged as an alternative for individually packaged items	
Organic Produce		Uses Nutri-Clean certification program to test for pesticide residues on produce	Limited number available
CFC Recovery	Policy encourages installing refrigerant recover systems	Installed recovery systems; operates preventive maintenance program	
Supplier Conditions	Encourages suppliers to package goods in reusable or recyclable materials	Encourages elimination of excess packaging, harmful chemicals, solvents, and gases; also promotes recycling codes	Adopted packaging reduction plan initiated by N.J. Food Council
Internal Recycling	Cardboard, plastic shrink wrap. All Kroger-brand egg cartons made of 100% recycled paperboard	Cardboard, computer paper, office paper, scrap paper, magazines, 35mm film holders, photo processing chemicals, plastic milk jugs	Cardboard, computer paper, office paper, newspapers, glass, aluminum
Energy Conservation	Recaps tires on vehicles; uses old tires as bumper guards; recycles motor oil	Energy management system in place; upgrades during remodels	Encourages conservation in all forms; company cafeteria gives discount on coffee for those who bring own mug

	Pratt Foods	Purity Supreme	Safeway
Company Policy	No formal policy	Includes statement, "We will make responsible environmental judgments"	Formal written policy
Consumer Information	Offers pamphlets on variety of topics. Advertises extensively on green topics; "Earth Bus" visits local schools	With Mass. Food Association, distributes brochure with environmental information	"Environmental Options" program includes shelf signs, brochures, and other information
Shelf Labeling	"Friendly Item—Oklahoma Green" shelf tags. Store posters encourage "4 R's"		Shelf labels explain benefits of products, based only on manufacturers' claims
Shopping Bags	Paper bags made of recycled content. 5-cent rebate for reuse or return. Sells canvas bags	Offers paper and plastic, both with recycled content; sells canvas and string bags. Donates 2 cents to charity for each returned bag	Offers both paper and plastic; most stores accept both for recycling. Sells inexpensive cloth bags
Recycling Drop-off	Aluminum, newspaper, glass, plastic soda bottles, milk jugs		Plastic, glass, aluminum in over half of stores' parking lots
Meat Department	Choice between paper and plasic wrapping; uses meat trays made of wood pulp		
Bulk Purchases	Encourages bulk buying through pamphlets and ads	Has Heartland Food Warehouses, catering to bulk buying	Available in some stores
Organic Produce	Wide variety offered. Displays read, "Our Veggies Don't Do Drugs"	Available for more than half of all produce selections	Only where consumer interest prevails
CFC Recovery			
Supplier Conditions		Contacted all suppliers requesting source reduction and recycled and recyclable packaging materials	"Environmental Options" requires suppliers to meet reduced or recycled packaging requirements
Internal Recycling	Office paper	Cardboard, office paper; donates day-old baked goods to local charities	Plastic, glass, aluminum, paper, cardboard, and oil and batteries from company trucks
Energy Conservation		Systems for efficient lighting, refrigeration, and air conditioning	Measures in place to save energy and water

	Shaw's	Spartan Stores	Winn-Dixie
Company Policy	Formal written policy	Formal written policy	Formal written policy
Consumer Information	Offers in-store brochures	Provides brochures on local recycling programs; sponsors consumer information programs	
Shelf Labeling		Some stores label products that are recyclable locally	
Shopping Bags	Offers paper and plastic made from some recycled contents; sells cloth bags. 5-cent rebate for reused paper bags	Offers paper and plastic; sells cloth bags. All paper bags contain recycled material; 3 to 5-cent rebates for returned paper bags	Offers paper and plastic, both made of partially recycled contents; sells cloth bags
Recycling Drop-off	Aluminum, paper, some plastics	Various stores accept plastic, glass, steel, newspaper, polystyrene	
Meat Department			
Bulk Purchases			Offers case-lot sales in all stores
Organic Produce	Some offered	Available in some stores	Offered when available in sufficient quantities at competitive prices
CFC Recovery	Uses recovery systems in all stores	Uses coolant recovery in fleet maintenance systems; computerized air-handling system eliminates some air conditioning needs	Uses recovery systems in all stores
Supplier Conditions		Works with suppliers to reduce packaging. Encourages labeling of recyclable plastics	
Internal Recycling	Computer and office paper	Office paper, waste ink, motor oil, Freon, cardboard, batteries, shrink wrap	Cardboard, plastic wrap, office paper
Energy Conservation	Various conservation programs in place	Computers plot delivery routes to save fuel; electronic engines increase fuel economy	Closely monitors water and energy use; awards given for conservation ideas

Supermarket Names and Addresses

A&P, *See Great Atlantic & Pacific Tea Co.*

Albertson's Inc.
250 Parkcenter Blvd.
Boise, ID 83706
208-385-6200

Angelo's Supermarkets, *See Supermarkets General Holdings Corp.*

Bells Supermarkets, *See Peter J. Schmitt Co., Inc.*

Big Bear
5075 Federal Blvd.
San Diego, CA 92102
619-263-3161

Big Star, *See Grand Union Co.*

Bi-Lo, Inc.
P.O. Drawer 99
Mauldin, SC 29662
803-234-1600

Bread & Circus
1163 Walnut St.
Newton, MA 02161
617-332-2400

Bruno's Inc.
P.O. Box 2486
Birmingham, AL 35201
205-940-9400

Dominick's Finer Foods, Inc.
505 Railroad Ave.
Northlake, IL 60164
708-562-1000

Dominions, *See Great Atlantic & Pacific Tea Co.*

Edwards, *See First National Supermarkets, Inc.*

Finast, *See First National Supermarkets, Inc.*

First National Supermarkets, Inc.
17000 Rockside Rd.
Maple Heights, OH 44137
216-587-7100

Food Emporium, *See Great Atlantic & Pacific Tea Co.*

Food Lion, Inc.
P.O. Box 1330
Salisbury, NC 28145
704-633-8250

Freshmart, *See Peter J. Schmitt Co., Inc.*

Giant Food Inc.
Box 1804
Washington, DC 20013
301-341-4582

Giant Eagle, Inc.
101 Kappa Dr.
Pittsburgh, PA 15238
412-963-6200

Grand Union Co.
201 Willowbrook Blvd.
Wayne, NJ 07470
201-890-6000

Great Atlantic & Pacific Tea Co.
2 Paragon Dr.
Montvale, NJ 07645
201-573-9700

Heartland, *See Supermarkets General Holdings Corp.*

Kohl's, *See Great Atlantic & Pacific Tea Co.*

Kroger Co.
1014 Vine St.
Cincinnatti, OH 45202
513-762-4000

Loblaws, *See Peter J. Schmitt Co., Inc.*

Fred Meyer, Inc.
P.O. Box 42121
Portland, OR 97242
503-232-8844

Miracle, *See Great Atlantic & Pacific Tea Co.*

Pathmark, *See Supermarkets General Holdings Corp.*

Pick-n- Pay, *See First National Supermarkets, Inc.*

Pratt Foods
130 S. Kickapoo St.
Shawnee, OK 74802
405-275-9831

Purity Supreme, *See Supermarkets General Holdings Corp.*

Safeway Stores, Inc.
4th St. & Jackson St.
Oakland, CA 94660
415-891-3000

Peter J. Schmitt Co., Inc.
355 Harlem Rd.
West Seneca, NY 14224-1892
716-825-1111

Shaw's Supermarkets, Inc.
140 Laurel St.
East Bridgewater, MA 02333
508-378-7211

Spartan Stores Inc.
850 76th St., SW
Grand Rapids, MI 49518
616-878-2000

Super Fresh, *See Great Atlantic & Pacific Tea Co.*

Supermarkets General Holdings Corp.
301 Blair Road
Woodbridge, NJ 07095
201-499-3000

Waldbaum, *See Great Atlantic & Pacific Tea Co.*

Winn-Dixie Stores, Inc.
P.O. Box B
Jacksonville, FL 32203
904-783-5000

Are These Guys Hiding Something?

Here are the names and addresses of companies that would not provide us with information about their environmental policies and practices. Do they have something to hide? We'll never know unless they open up. If you shop at one of these chains, perhaps you can persuade them to let their customers know what, if anything, they are doing to help the environment. Give the store manager where you shop a copy of the questionnaire on page 166 and ask for answers. Send a copy to the corporate offices. (We've provided the name of each company's chief executive officer.) Tell them that you want answers.

Let us know what you find out. We'll share the findings with other readers.

Acme Markets, *See American Stores Co.*

Alpha Beta, *See American Stores Co.*

American Stores Co.
P.O. Box 27447
Salt Lake City, UT 84127
801-539-0112
Jonathan L. Scott, President

Bradley's, *See Stop & Shop Companies*

H.E. Butt Grocery Co.
4839 Space Center Dr.
San Antonio, TX 78218
512-662-5414
Charles Butt, President

Hy-Vee Food Stores, Inc.
1801 Osceola Ave.
Chariton, IA 50049
515-774-2121
Ronald Pearson, President

[continued on next page]

How to Rate Your Supermarket

We've provided a form on page 166 you can use to compare supermarkets in your area, to determine which is greenest. Feel free to make a copy of it so that it will be easier to use.

Before filling it out, it may be helpful to review the twelve criteria on pages 152-153. When you do your survey, you should talk with the manager of the store you are rating. Better yet, contact the store's corporate headquarters (if it is a chain) and speak to their community relations or consumer information office. (This may be an ideal project for a classroom, a scout troop, or other kid's organization.) You may choose to rate each store however you'd like: simple "yes" and "no" answers,

Jewel Food Stores, *See American Stores Co.*

Lucky, *See American Stores Co.*

Publix Super Markets, Inc.
1936 George Jenkins Blvd.
Lakeland, FL 33801
813-680-5250
Howard Jenkins, Chairman

Ralph's Grocery Co.
2215 S. Wilmington
Compton, CA 90220
213-637-1101
Byron Allumbaugh, CEO

Scaggs Alpha Beta, *See American Stores Co.*

Smith's Food and Drug, Inc.
P.O. Box 527
Layton, UT 84041
801-974-1400
Richie Smith, President

Star Market, *See American Stores Co.*

Stop & Shop Cos., Inc.
P.O. Box 1942
Boston, MA 02105
617-380-8000
Lewis Schaeneman, President

Super Value Stores, Inc.
11840 Valley View Rd.
Eden Prairie, MN 55344-3691
612-828-4000
Michael Wright, CEO

Vons Cos., Inc.
618 Michillinda Ave.
Arcadia, CA 91007
818-821-7000
Roger Stangeland, CEO

school grades (A, B, C, D, and F), or using a 1-to-10 scale. You may even want to devise your own form with enough space to provide more detailed answers.

Demand the answers you are looking for. Be polite but firm. If a store's management won't answer your questions to your satisfaction, shop elsewhere!

Share your findings with your friends, neighbors, church group, or community association. (If you make a comprehensive comparison of all the supermarkets in your area, consider notifying a local newspaper or television station of your findings.) And share your findings with us, so that we can include them in future editions of this book. Write to us c/o The Green Consumer, 1526 Connecticut Ave. NW, Washington, DC 20036.

The Green Supermarket Scorecard

In rating stores, use "yes" or "no" answers, grades (A,B, C, D, and F), or rate each on a scale of 1 to 10.

Subject	STORE		
	A	B	C
Environmental Policy (Do they have one?)			
Consumer Information (Do they offer help on being a Green Consumer?)			
Shelf Labeling (Are green products clearly marked?)			
Shopping Bags (Do they offer a choice and accept bags for recycling?)			
Recycling Facilities (Do they provide drop-off space for customers?)			
Meat Department (Do they offer paper wrapping and minimize waste?)			
Bulk Purchases (Do they make these easy and affordable?)			
Organic Produce (Is it offered for sale and clearly marked?)			
CFC Recovery (Do they actively protect the ozone layer?)			
Supplier Conditions (Must suppliers meet "green" standards?)			
Internal Recycling (Do they recycle in-house?)			
Energy Conservation (Do they actively conserve resources?)			

Part Four

A·C·T·I·O·N GUIDE

ACTION GUIDE

We can't overemphasize how important it is to make your voice heard. It's really easy. And it can have an impact.

We hear the same thing from corporate executives, legislators, supermarket owners, and just about anyone else who's "in charge": It only takes a few letters before they'll begin to pay attention.

Consider this: Wal-Mart, one of the top three retailers in the United States, sees approximately 70 million customers come through its doors *every week*. And yet, they say, as few as 20 letters on a given subject will cause them to take notice and contemplate action.

Twenty letters out of seventy million customers!

We've included the addresses of most of the nation's major companies that manufacture food or groceries sold in supermarkets. We urge you to write with your compliments and complaints, and to let them know that you are concerned about how their products affect the environment. If you purchased their product because it was "greener" than others, let them know. If you stopped buying their product because it was not environmentally responsible, let them know that, too. If you have questions about the environmental impact of their product, call or write and ask.

You can send a message—one purchase at a time.

AAGC/Birds Eye, *See Philip Morris Cos.*

After the Fall Products Inc.
Box 777
Brattleboro, VT 05301
802-257-4616

Ak-Mak Bakeries
89 Academy Ave.
Sanger, CA 93657
209-875-5511

Alberto Culver Co.
2525 Armitage Ave.
Melrose Park, IL 60160
312-450-3000; 800-822-2327

Allen's Naturally
P.O. Box 514
Farmington, MI 332514
313-453-5410

Allied Foods Inc.
1970 Hills Ave. NW
Atlanta, GA 31328
404-351-2400

Almay, *See Revlon Inc.*

Alpo Pet Foods Inc.
Grand Metropolitan USA
P.O. Box 2187
Allentown, PA 18001
215-395-3301; 800-366-6033

American Cyanamid Co.
One Cynanamid Plaza
Wayne, NJ 07470
201-340-6000

American Home Foods, *See American Home Products Corp.*

American Home Products Corp.
685 Third Ave.
New York, NY 10017
212-878-5000

Apple & Eve Inc.
P.O. Box 2137
Great Neck, NY 11022
516-829-6881

Aramis, *See Estee Lauder*

Arrowhead Mills Inc.
Box 2059
Hereford, TX 79045
806-364-0730

Ashdun Industries
400 Sylvan Ave.
Englewood Cliffs, NJ 07632
201-569-3600; 800-447-3008 ext.643

Aveda Corp.
400 Central Ave. SE
Minneapolis, MN 55414
612-378-7420; 800-448-6265

Avon
9 W. 57th St.
New York, NY 10019
212-546-6015; 800-858-8000

Baby Bunz & Co.
P.O. Box 1717
Sebastopol, CA 95473
707-829-5347

Barbara's Bakery Inc.
3900 Cypress Dr.
Petaluma, CA 94952
707-165-2273

Bare Essentials
809 University Ave.
Los Gatos, CA 95030
408-354-8853; 800-227-3386

Beatrice Cheese Inc.
ConAgra Co.
Two N. La Salle St.
Chicago, IL 60602
414-782-2750

Beatrice/Hunt-Wesson Inc.
ConAgra Co.
1645 W. Valencia Dr.
Fullerton, CA 92634
714-680-1000

B.G. Shrimp Sales Co.
Katy Industries Inc.
Box 5009
Tampa, FL 33675
813-248-4965

Biobottoms
P.O. Box 1060
Petaluma, CA 94953
707-778-7945

Booth Food Products
National Sea Products Inc.
79 Mechanic St.
Rockland, ME 04841
207-594-7860

Borden Inc.
277 Park Ave.
New York, NY 10172
212-573-4000; 800-426-7336

Boston Tea Co.
520 Secaucus Rd.
Secaucus, NJ 07094
201-865-0200

Boyle Midway Household Prods., *See American Home Products Corp.*

Bremner Inc., *See Ralston Purina Co.*

Brilliant Seafood Co. Inc.
315 Northern Ave.
Boston, MA 02210
617-338-0300

Bristol-Myers Squibb Co.
345 Park Ave.
New York, NY 10154
212-546-4000

Buitoni Foods Corp.
450 Huyler St.
South Hackensack, NJ 07606-1598
201-440-8000

Bumble Bee Seafoods Inc.
P.O. Box 23508
San Diego, CA 92123
619-560-0404; 800-767-4466

Cadbury Schweppes Inc.
P.O. Box 3800
Stamford, CT 06905
203-329-0911; 800-543-7276

Caliente Chile Inc.
P.O. Drawer 5340
Austin, TX 78763
512-472-6996

Campbell Soup Co.
Campbell Pl.
Camden, NJ 08103-1799
609-342-4800; 800-257-8443

Canada Dry/Sunkist USA, *See Cadbury Schweppes Inc.*

Caprisun Inc.
1825 S. Grant St.
San Mateo, CA 94402
415-572-7100

Carnation Co., *See Nestle Enterprises Inc.*

Carol Hansen, *See S. C. Johnson & Son Inc.*

Carter Products
Carter-Wallace Inc.
Half Acre Rd.
Cranbury, NJ 08512
609-655-6000

Celestial Seasonings Inc.
1780 55th St.
Boulder, CO 80301
303-449-3779

Charles of the Ritz, *See Revlon Inc.*

Chempoint Products Co.
HPM Corp.
543 Tarrytown Rd.
White Plains, NY 10607
914-279-2613; 800-343-6588

Chesebrough-Pond's Inc., *See Unilever USA*

Christian Dior Perfumes
9 W. 57th St.
New York, NY 10019
212-759-1840

Church & Dwight Co.
469 N. Harrison St.
Princeton, NJ 08543-5297
609-683-5900

Clairol, *See Bristol-Myers Squibb Co.*

Clinique, *See Estee Lauder*

Clorox Co.
1221 Broadway
Oakland, CA 94612
415-271-7000; 800-227-1860

Coca-Cola Co.
One Coca-Cola Plaza N.W.
Atlanta, GA 30313
404-676-2121; 800-438-2653

Colgate-Palmolive Co.
300 Park Ave.
New York, NY 10022
212-310-2000; 800-338-8388

Comstock Michigan Fruit Div.
Curtice-Burns Foods Inc.
248 9th St., P.O. Box 68
Benton Harbor, MI 49022-4723
616-927-4411

Conagra Consumer Frozen Foods Co.
ConAgra Co.
13515 Main St.
Omaha, NE 68102
402-595-6000; 800-722-1344

Conair Corp.
1 Cummings Point Rd.
Stamford, CT 06904
203-351-9000

Confab Sales & Marketing
175 Strafford Ave.
Wayne, PA 19087
800-262-0042

Conners Bros. Limited
Main St.
Blacks Harbor, NB E0G 1H0, Canada
506-456-3391

Continental Baking Co., *See Ralston Purina Co.*

Cosmair Inc., *See Nestle Enterprise Inc.*

CPC International Inc.
International Plaza, P.O. Box 8000
Englewood Cliffs, NJ 07632
201-894-4000

Crabtree & Evelyn Ltd.
P.O. Box 167
Woodstock, CT 06281
203-928-2761

Crosse & Blackwell Co., *See Nestle Enterprises Inc.*

Cumberland Packing Corp.
2 Cumberland St.
Brooklyn, NY 11205
718-858-4200

Dare Foods Ltd.
2481 Kingsway Dr.
Kitchener, ON
N2G 4G4, Canada
519-893-5500

Dannon Co. Inc.
1111 Westchester Ave.
White Plains, NY 10604
914-697-9700

Jimmy Dean Meat Co. Inc., *See Sara Lee Corp.*

Deer Park Spring Water Inc., *See Clorox Co.*

Del Monte USA, *See RJR Nabisco Inc.*

Devonsheer, *See CPC International Inc.*

Dial Corp.
Greyhound Dial Corp.
111 W. Clarendon Ave.
Phoenix, AZ 85077
602-248-4000

Dole Food Co.
Castle & Cooke Inc.
10900 Wilshire Blvd., Ste. 1500
Los Angeles, CA 90024
213-824-1500; 800-232-8888

Dolefam Corp.
2000 K St.
Washington, DC 20006
202-775-9715

DowBrands Co., *See Dow Chemical Co.*

Dow Chemical Co.
2030 Willard H. Dow Center
Midland, MI 48674
517-636-1000

Drackett Co., *See Bristol-Myers Squibb Co.*

Droste USA Ltd.
Park 80 W. Plaza 1
Saddlebrook, NJ 07662
201-712-1800

Duncan Hines, *See Procter & Gamble Co.*

Dutch Gold Honey Inc.
2220 Dutch Gold Dr.
Lancaster, PA 17601
717-393-1716

Eagle Snacks Inc.
One Busch Pl.
St. Louis, MO 63118
314-966-7100

Earth Wise Inc.
1790 30th St.
Boulder, CO 80301
303-447-0119

Earth's Best Inc.
P.O. Box 887
Middlebury, VT 05753
802-388-7974; 800-442-4221

EarthRite
Magic American Corp.
23700 Mercantile Rd.
Beechwood, OH 44122-5955
216-464-2353; 800-321-6330

Eastman Kodak Co.
343 State St.
Rochester, NY 14650
716-724-4000

Eat All Frozen Food Co. Inc.
Box 14020
Philadelphia, PA 19122
215-769-6800

Ecomer Inc.
13 N. 17th St.
Perkasie, PA 18944
215-453-2505

Ecoworks
2326 Pickwick Rd.
Baltimore, MD 21207
301-448-1820

Eden Foods Inc.
701 Tecumseh-Clinton Rd.
Clinton, MI 49236
517-456-7424

Edward Lowe Industries Inc.
348 S. Columbia St., P.O. Box 16
South Bend, IN 46624
219-288-6225

Entenmann's Inc., *See Philip Morris Cos.*

Estee Lauder
767 Fifth Ave.
New York, NY 10153
212-572-4200

Estee Corp.
169 Lackawanna Ave.
Parsippany, NJ 07054-1094
201-335-1000

Eveready Battery Co., *See Ralston Purina Co.*

Faberge USA Inc., *See Unilever USA*

Fantastic Foods Inc.
106 Galli Dr.
Novato, CA 94949
415-883-7718

Faultless Starch/Bon Ami Co.
1025 W. 8th St.
Kansas City, MO 64101-1200
816-842-1230

First Brands Corp.
P.O. Box 1901
Danbury, CT 06813-1911
203-731-2300

Fisher Nut Co.
Procter & Gamble Co.
2327 Wycliff St.
St. Paul, MN 55114
612-645-0635; 800-543-0480

Flex, *See Revlon Inc.*

Fort Howard Corp.
P.O. Box 19130
Green Bay, WI 54307
414-435-8821

Fox Run Craftsman
Ribble Corp.
1907 Stout Run
Warminster, PA 18974
215-675-7700

R.T. French Co.
Reckitt & Colman North America
One Mustard St.
Rochester, NY 19406
716-482-8000

Frito-Lay Inc., *See PepsiCo Inc.*

Fromageries Bel Inc.
P.O. Box 2109
Fort Lee, NJ 07024-8609
201-592-6601

R.W. Frookies Inc.
375 Sylvan Ave.
Englewood Cliffs, NJ 07632
201-871-3335

Galletti Brothers
2900 Ayers Ave.
Los Angeles, CA 90023
213-262-8222

General Biscuit of America Inc.
BSN Group
891 Newark Ave.
Elizabeth, NJ 07207
201-527-7000

General Foods, *See Kraft General Foods Inc.*

General Mills Inc.
P.O. Box 1113
Minneapolis, MN 55440
612-540-2311; 800-222-6846

Georgia-Pacific Corp.
133 Peachtree St. NE
Atlanta, GA 30303
404-521-4000

Gerber Products Co.
445 State St.
Boston, MA 02111
617-330-1800; 800-4GERBER

Gillette Co.
101 Huntington Ave.,
Boston, MA 02199
617-421-7000

Glamorene, Division of Airwick Industries Inc.
111 Commerce Rd.
Carlstadt, NJ 07072-2595
201-933-8200

Glenbrook Laboratories, *See Eastman Kodak Co.*

Glencourt Inc.
2800 Ygnacio Valley Rd.
Walnut Creek, CA 94598-3592
415-944-4395

Golden Dipt Co.
DCA Food Industries Inc.
12813 Flushing Meadows Dr.
St. Louis, MO 63131-1835
314-821-3113

Golden Grain Co., *See Quaker Oats Co.*

Gorton's of Gloucester, *See General Mills Inc.*

Goya Foods Inc.
100 Seaview Dr.
Secaucus, NJ 07094
201-348-4900

Hain Pure Food Co., *See Pet Inc./Whitman Corp.*

Health Valley Food Inc.
16100 Foothill Blvd.
Irwindale, CA 91706-7811
818-334-3241; 800-423-4846

Heinz Pet Products
180 E. Ocean Blvd.
Long Beach, CA 90802
213-590-8100

Helene Curtis Industries Inc.
325 N. Wells St.
Chicago, IL 60610
312-661-0222

Hershey Foods Corp.
19 E. Chocolate Ave.
Hershey, PA 17033-0815
717-534-4200; 800-468-1714

Hillshire Farm & Kahn Co., *See Sara Lee Corp.*

H.J. Heinz Co.
1062 Progress St.
Pittsburgh, PA 15252
412-237-5757

Hollywood Foods, *See Pet Inc./ Whitman Corp.*

Geo. A. Hormel & Co.
P.O. Box 800, 501 16th Ave. N.E.
Austin, MN 55912
507-437-5611; 800-523-4635

Humble Whole Foods
2500 Old Crow Canyon Rd., #218
San Ramon, CA 94583
415-837-8888; 713-446-8885

H.V.R. Co., *See Clorox Co.*

HyGeia Sciences Inc.
Tambrands Inc.
1 Marcus Ave.
Lake Success, NY 11042
516-358-8300

Integrity Baking Co.
RD 3, Box 788
Franklinville, NJ 08322
609-694-4235

James River Corp.
P.O. Box 2218
Richmond, VA 23217
804-644-5411

Jean Nate, *See Revlon Inc.*

Johanna Farms Inc.
Johanna Farms Rd.
Flemington, NJ 08822-0000
201-788-2200

Johnson & Johnson
CPI Information Center
199 Grandview Rd.
Skillman, NJ 08858
201-874-1000; 800-526-3967

S.C. Johnson & Son Inc.
1525 Howe St.
Racine, WI 53403
414-631-2000; 800-558-5252

Kal Kan Foods Inc.
3250 E. 44th St.
Vernon, CA 90058
213-587-2727; 800-KAL KARE

Keebler Co.
United Biscuits Holdings PLC
One Hollow Tree Lane
Elmhurst, IL 60126
312-833-2900

Kellogg Co.
One Kellogg Sq.
Battle Creek, MI 49016
616-961-2000

Kimberly-Clark Corp.
545 E. John W. Carpenter Fwy., Ste. 1300
Irving, TX 75062-3968
214-830-1200

Kiss My Face Corp.
P.O. Box 804
New Paltz, NY 12561
914-255-0884

Kiwi Brands, *See Sara Lee Corp.*

Knouse Foods Co-Operative Inc.
800 Peach Glen Rd.
Peach Glen, PA 17306
717-667-8181

R.W. Knudsen & Sons Inc., *See J.M. Smucker Co.*

Kosher Empire Poultry Inc.
River Rd., Box 165
Mifflintown, PA 17059
800-367-4734

Kraft General Foods Inc.
Philip Morris Cos.
250 North St.
White Plains, NY 10625
914-335-5849; 800-431-1001

Krusteaz
Continental Mills Inc.
P.O. Box 88176
Seattle, WA 98138
206-872-8400

Land O'Lakes Inc.
P.O. Box 116
Minneapolis, MN 55126
612-481-2222

Anne Lanyi Foods
Cherry Hill Cooperative Cannery
P.O. Box 2032
South Londonberry, VT 05155
802-824-6275

Lawry's Foods Inc., *See Unilever USA*

Lehn & Fink Products Group, *See Eastman Kodak Co.*

Lever Brothers Co., *See Unilever USA*

Life Tree Products
Sierra Dawn Products
P.O. Box 513
Graton, CA 95444
707-577-0205

J.R. Liggett Ltd.
Rte. 12A, RR2, Box 911
Cornish, NH 03745
603-675-2055

Thomas J. Lipton Co., *See Unilever USA*

Little Bear Organic Foods
15200 Sunset Blvd., Ste. 202
Pacific Palisades, CA 90272
213-454-4542

Livos PlantChemistry
1365 Rufina Circle
Santa Fe, NM 87501
505-438-3448; 800-621-2591

Lloyd's Food Products Inc.
1455 Mendota Heights Rd.
St. Paul, MN 55120
612-688-6000

Loma Linda Foods Inc.
11503 Pierce St.
Riverside, CA 92505-3319
714-687-7800

M&M/Mars Inc.
High St.
Hackettstown, NJ 07840
201-852-1000

Marcal Paper Mills Inc.
1 Market St.
Elmwood Park, NJ 07407
201-796-4000

Maruchan Inc.
1902 Deere Ave.
Irvine, CA 92714
714-261-6525

Matlaw's Food Products Inc.
135 Front Ave.
West Haven, CT 06516
203-934-5233

Mauna Loa Macadamia Nut Corp.
A.C. Breur Co.
HC1, Box 3
Hilo, HI 96720
808-966-9301

Max Factor, *See Revlon Inc.*

Mayacamas Fine Foods Inc.
1206 E. McArthur
Sonoma, CA 95476
707-996-0955

M.C. Snack Inc.
P.O. Box 84329
San Diego, CA 92138
619-484-2170

McCadam Cheese Co. Inc.
Dean Foods
P.O. Box 345
Heuvelton, NY 13654
315-344-2441

McCain Citrus Inc.
1821 S. Kilbourn
Chicago, IL 60623-2391
312-762-9000

McCain Elios's Foods Inc.
11 Gregg St.
Lodi, NJ 07644
201-368-0600

McCormick & Co.
11350 McCormick Rd., P.O. Box 208
Hunt Valley, MD 21031-0208
301-771-7301

Mead Johnson Nutritionals, *See Bristol-Myers Squibb Co.*

Melitta North America Inc.
1401 Berlin Rd.
Cherry Hill, NJ 08003
609-428-7202

Mennen Co.
Hanover Ave.
Morristown, NJ 07962-1928
201-631-9000

Mercantile Foods Co.
P.O. Box 1140
Georgetown, CT 06829
203-544-9891

Millflow Spice Corp.
105-25 180th St.
Jamaica, NY 11433
718-657-4755

Mobil Corp.
1169 Pittsford-Victor Rd.
Pittsford, NY 10017
716-248-1246

Modern Products Inc.
3015 W. Vera Ave.
Milwaukee, WI 53209
414-352-3208

Moore's Food Products, *See Clorox Co.*

Morton Salt Co.
110 N. Wacker Dr.
Chicago, IL 60606-1555
312-807-2643

Mott's USA, *See Cadbury Schweppes Inc.*

Mountain Fresh
P.O. Box 30416
Grand Junction, CO 81504
303-434-8434

Mrs. Paul's Kitchens Inc., *See Campbell Soup Co.*

Mrs. Smith's Frozen Foods Co., *See Kellogg Co.*

Mui Li Wan Inc.
2300 Nevada Ave.
North Golden Valley, MN 55427
612-542-1805

Murphy-Phoenix Co.
Box 22930
Cleveland, OH 44122
216-831-0404

Nabisco Brands Inc., *See RJR Nabisco Inc.*

Natural Inc.
6650 Santa Barbara Rd.
Elkridge, MD 21227
301-621-5388

Natural Chemistry Inc.
224 Elm St.
New Canaan, CT 06840
203-966-8761; 800-753-1233

Nestle Enterprises Inc.
3003 Bainbridge Rd.
Solon, OH 44139
216-349-5757; 800-NESTLES

Nestle Food Corp., *See Nestle Enterprises Inc.*

Neutrogena Corp.
5760 W. 96th St., P.O. Box 45036
Los Angeles, CA 90045
213-642-1150

Newman's Own Inc.
246 Post Rd. E.
Westport, CT 06880
203-222-0136

Nissin Foods USA Co. Inc.
2001 W. Rosecrans Ave.
Gardena, CA 90249-2994
213-321-6453

Noxell Corp., *See Procter & Gamble Co.*

Ocean Spray Cranberries Inc.
225 Water St.
Leominister, MA 01453
508-946-1000

Old Fashioned Kitchen Inc.
1045 Towbin Ave.
Lakewood, NJ 08701-5931
201-364-4100

Ore-Ida Foods Co. Inc., *See H.J. Heinz Co.*

Oscar Mayer Foods Inc., *See Philip Morris Cos.*

Pasta Valente
Box 2307
Charlottesville, VA 22902
804-971-3717

Pepperidge Farm Inc., *See Campbell Soup Co.*

PepsiCo Inc.
P.O. Box 442
Somers, NY 10589
914-767-6000; 800-352-4477

Pet Inc./Whitman Corp.
400 South Fourth St.
St. Louis, MO 63109
314-621-5400; 800-842-3442

Pfizer Inc.
235 E. 42nd St.
New York, NY 10017
212-573-2323

Philip Morris Cos.
120 Park Ave.
New York, NY 10017
212-880-5000; 800-431-1001

Pillsbury Co.
Grand Metropolitan USA
2866 Pillsbury Center
Minneapolis, MN 55402
612-330-4966; 800-767-4466

Planters LifeSavers Co., *See RJR Nabisco Inc.*

Playtex Inc.
183 Madison Ave., Ste. 1600
New York, NY 10016
212-532-1700

M. Polaner & Sons Inc.
426 Eagle Rock Ave.
Roseland, NJ 07068
201-228-2500

Power Plus of America
1605 Lakes Pkwy
Lawrenceville, GA 30243
404-339-1672

Procter & Gamble Co.
P.O. Box 599
Cincinnati, OH 45202
513-983-1100; 800-543-0480,-1745

Progresso Quality Foods Corp., *See Pet Inc./Whitman Corp.*

Quail Oval Farms, *See RJR Nabisco Inc.*

Quaker Oats Co.
Quaker Tower, 321 N. Clark St.
Chicago, IL 60610
312-222-7111; 800-558-5252

Ralston Purina Co.
Checkerboard Square
St. Louis, MO 63164
314-982-1000; 800-345-5678

Red L. Foods Inc.
15 Commerce Dr.
Monroe, CT 06468
203-268-8633

Red Oak Farms, *See RJR Nabisco*

Reese Finer Foods, *See Pet Inc./Whitman Corp.*

Revlon Inc.
767 Fifth Ave.
New York, NY 10153
212-572-5000

Reynolds Metals Co.
6601 W. Broad St.
Richmond, VA 23230-1701
804-281-2000

Rich Products Corp.
1145 Niagara St.
Buffalo, NY 14213
716-878-8000

Rich-Seapak Corp.
P.O. Box 667
Saint Simons Island, GA 31522-0000
912-638-5000

RJR Nabisco Inc.
9 W. 57th St.
New York, NY 10019
212-258-5600

Rockline Inc.
Box 1007
Sheboygan, WI 53082
414-459-4160

Ross Laboratories
Abbott Laboratories
625 Cleveland Ave.
Columbus, OH 43215
614-227-3333

Royal Crown Beverage Co.
553 N. Fairview
Saint Paul, MN 55104
612-645-0501

Salad King Inc., *See Newman's Own Inc.*

Sandoz Nutrition Corp.
5100 Gambel Dr., P.O. Box 370
Minneapolis, MN 55416-1547
612-925-2100

Santa Cruz Natural, *See J.M. Smucker Co.*

Sanwa Foods Inc.
331 N. Vineland Ave.
City of Industry, CA 91146
818-369-5041

Sara Lee Corp.
500 Waukegun Rd.
Deerfield, IL 60015
312-945-2525; 800-323-7117

Schering-Plough Corp.
One Giralda Farms
Madison, NJ 07940-1000
201-822-7000

Schweppes USA, *See Cadbury Schweppes Inc.*

Scott Paper Co.
Scott Plaza
Philadelphia, PA 19113
215-522-5000; 800-TEL-SCOT

Seagram & Sons, Joseph E. Inc.
375 Park Ave.
New York, NY 10152
212-572-7000

Sego, *See Pet Inc./Whitman Corp.*

Seneca Foods Corp., *See Nestle Enterprises Inc.*

Seven Up Bottling Co.
P.O. Box 655086
Dallas, TX 75265
214-360-7000

Shaffer & Clarke & Co. Inc.
3 Parklands Dr.
Darian, CT 06820-3639
203-655-3555

Shedds Foods Inc., *See Unilever USA*

Shulton Inc. USA, *See American Cyanamid Co.*

Simply Pure Baby Food
RFD 3, Box 99
Bangor, ME 04401
207-941-1924; 800-426-7873

Sioux Honey Co-Op
509 Lewis Blvd.
Sioux City, IA 51105
712-258-0638

Sisterly Works
RR 3 Box107
Port Lavaca, TX 77979
512-893-5252

Smartfoods Inc., *See PepsiCo Inc.*

J.M. Smucker Co.
1 Strawberry Ln.
Orrville, OH 44667-1298
216-682-0015

Snyder's of Hanover Inc.
Curtice-Burns Foods Inc.
P.O. Box 471
Hanover, PA 17331
717-632-4477

Special Products, *See CPC International Inc.*

Specialty Brands Inc.
222 Sutter St.
San Francisco, CA 94108-4496
415-981-7600

StarKist SeaFood Co., *See Heinz Pet Products*

Stinson Canning Co.
Rt. 186
Prospects Harbour, ME 04669
207-963-7331

Stokely-Van Camp Inc., *See Quaker Oats Co.*

Stonyfield Farms, Inc.
10 Burton Dr.
Londonderry, NH 03053
603-437-4040

Stouffer Foods Corp., *See Nestle Enterprises Inc.*

Stretch Island Fruit Inc.
E. 370 Greenway
Grapeview, WA 98546
206-275-6050

Sun-Diamond Growers of California
5568 Gibralter
Pleasanton, CA 94566-4053
415-463-8200

Sundor Brands Inc., *See Procter & Gamble Co.*

Sunshine Food Markets Inc.
P.O. Box 1108
Sioux Falls, SD 57117
605-336-2505

Sunshine Makers
15922 Pacific Coast Highway
Huntington Harbor, CA 92649
213-592-2844; 800-228-0709

Superior Brands Inc.
122 Quincy Shore Dr., P.O. Box 89
Quincy, MA 02171-2993
617-770-0880

Swift/Eckrich
ConAgra Co.
2001 Butterfield Rd.
Downers Grove, IL 60515
708-574-7015; 800-ECK-RICH

Tetley Inc.
100 Commerce Dr.
Shelton, CT 06484
203-929-9200

S.B. Thomas Inc., *See CPC International Inc.*

Tom's of Maine
Railroad Ave.
Kennebunk, ME 04043
207-985-2944

Traditionals Inc.
4515 Ross Rd.
Sebastapol, CA 95472
707-823-8911

Tropicana Products Inc., *See Seagram & Sons, Joseph E. Inc.*

Tyson Foods Inc.
Holly Farms Corp.
2210 W. Oaklawn Dr.
Springdale, AR 72765
501-756-4000; 800-233-6332

Uncle Ben's Inc.
Mars Inc.
5721 Harvey Wilson Dr.
Houston, TX 77251
713-674-9484

Unilever USA
390 Park Ave.
New York, NY 10022
212-688-6000; 800-243-5804

Union Camp Container Corp.
1 Edmund Rd., P.O. Box 498
Newton, CT 06470-0498
203-426-5871

Upjohn Co.
7000 Portage Rd.
Kalamazoo, MI 49001
616-323-4000

Van De Kamp's Frozen Foods, *See Pet Inc./Whitman Corp.*

Van Den Bergh Foods Co., *See Unilever USA*

Veryfine Products
P.O. Box 2425, Harwood Station
Littleton, MA 01460
508-486-3521

Vlasic Foods Inc., *See Campbell Soup Co.*

Waverly Mineral Products Co. Inc.
555 City Line Ave.
Bala-Cynwyd, PA 19004-1104
215-668-2808

Webster Industries
58 Pulaski St., P.O. Box 3119
Peabody, MA 01960
508-532-2000

Welch Foods Inc.
100 Main St.
Concord, MA 01742
508-371-1000

Wella Corp.
524 Grand Ave.
Englewood, NJ 07631
201-569-1020

Westbrae Natural Foods Inc.
P.O. Box 91-1181
Commerce, CA 90091
213-722-1692

White Castle Distributing Inc.
555 W. Goodale St.
Columbus, OH 43215
614-228-5781

Whitehall Laboratories Inc., *See American Home Products Corp.*

Wonder Chemical Corp.
P.O. Box 56
Fairless Hills, PA 19030
215-945-4300

Worthington Foods Inc.
900 Proprietors Rd.
Worthington, OH 43085-3194
614-885-9511

Wyeth-Ayerst Laboratories, *See American Home Products Corp.*

Yoplait USA Inc., *See General Mills Inc.*

Shopping by Mail

Here is a select list of firms that sell green products by mail. Write or call for catalogs.

AFM Enterprises
1140 Stacey Ct.
Riverside, CA 92507
714-781-6860

Allen's Naturally
P.O. Box 339
Farmington, MI 48332
313-453-5410

The Body Shop Inc.
45 Horsehill Rd.
Cedar Knolls, NJ 07927
800-541-2535

Co-Op America
2100 M St. NW
Washington, DC 20036
202-223-1881

Earth Care Paper Inc.
P.O. Box 7070
Madison, WI 53707
608-277-2900

Ecco Bella
6 Provostt Sq., Ste. 602
Caldwell, NJ 07006
800-888-5320

Gardener's Supply
128 Intervale Rd.
Burlington, VT 05401
802-863-1700

Jade Mountain
P.O. Box 4616
Boulder, CO 80306
303-449-6601

Livos PlantChemistry
614 Aqua Fria St.
Santa Fe, NM 87501
505-988-9111

The Natural Baby Company
RD 1, Box 160S
Titusville, NJ 08560
609-737-2895

Necessary Trading Co.
New Castle, VA 24127
703-864-5103

People for the Ethical Treatment of Animals
P.O. Box 42516
Washington, DC 20015
301-770-7382

Real Goods
966 Mazzoni St.
Ukiah, CA 95482
800-762-7325

Seventh Generation
49 Hercules Dr.
Colchester, VT 05446
800-456-1177

INDEX

The Garden of Eden and the Hope of Immortality

The Garden of Eden and the Hope of Immortality

James Barr

FORTRESS PRESS MINNEAPOLIS

THE GARDEN OF EDEN AND THE HOPE OF IMMORTALITY

First Fortress Press edition published 1993.

Library of Congress Cataloging-in-Publication data
Barr, James, 1924–
The Garden of Eden and the hope of immortality / James Barr.
p. cm.
Includes bibliographical references and index.
ISBN 0–8006–2744–X
1. Eden. 2. Bible. O.T. Genesis III--Criticism, interpretation, etc. 3. Immortality--Biblical teaching. 4. Resurrection--Biblical teaching. I. Title.
BS1237.B37 1993
236′.22--dc20 92–27037
CIP

Manufactured in Great Britain 1–2744
97 96 95 94 93 1 2 3 4 5 6 7 8 9 10

This book is based on five lectures given in the University of Bristol in 1990, when the author was invited to be the Read-Tuckwell lecturer for that year. The Read-Tuckwell lectureship was established by a residual bequest to the University of Bristol made by Alice Read-Tuckwell, who directed in her will that income deriving from the trust funds should be used to establish and maintain the lectureship and that the lecturer should deliver a course of lectures on Human Immortality and related matters, the course of lectures to be printed and published.

Contents

Preface

This book is an expanded version of the fifth series of Read-Tuckwell Lectures, delivered at Bristol University in May 1990. The Lectures are specifically endowed for the study of the subject of human immortality. I was honoured by the invitation from Bristol University to deliver these lectures, and was pleased and stimulated by the opportunity they offered, an opportunity to explore certain connections which run through the entire Bible and have ramifications in numerous aspects of theology and ethics.

Central to the presentation offered here is a reading of the story of the Garden of Eden, not as a tale of the origins of sin and death, but as a tale of a chance of immortality, briefly accessible to humanity but quickly lost. Old Testament scholars have long known that the reading of the story as the 'Fall of Man' in the traditional sense, though hallowed by St Paul's use of it, cannot stand up to examination through a close reading of the Genesis text. But though this has long been evident, scholars have not, on the whole, succeeded in formulating a general picture of the purpose and impact of the story which could rival the traditional one and could carry an equal force or similar relevance over so wide a range of biblical materials and theological considerations. My suggestion is that the perception of the place of immortality in the narrative can do much to fulfil that need.

The subject is one that well illustrates the question of the relation between the Old Testament and the New. It has been, above all, Paul's own reading of the story that has led people, all down the centuries, to read it in a certain way. Paul's reading itself, however, depends heavily on a particular tradition of interpretation which is not always evident to the ordinary reader. If one is to reconsider the ways in which the Old and New Testaments may mutually illuminate each other within Christianity, no example is more profound or more far-reaching than the story of Adam and Eve.

The Read-Tuckwell endowment, by fixing interest on the subject of human immortality, rightly made central an aspect which down the ages has always or normally been regarded as central and essential to religion. The twentieth century, however, has left people confused in this regard, for powerful voices have been heard to declare that the focus of Christianity is entirely directed upon the resurrection of the body, and that the immortality of the soul, which is the commonest formulation of human immortality, is a quite alien idea, based upon Greek philosophy and foreign to the Hebrew heritage of ideas upon which Christian faith rests. The study here presented may, by indicating a path through which Hebrew thinking also was concerned with immortality, help to overcome some of this uncertainty.

Our study may also have something to say about the relation between biblical texts and studies in other areas: particularly three, namely (1) philosophy, (2) anthropology, and (3) the general history of religions. (1) A large part of the traditional argumentation concerning immortality was always philosophical: more modern trends in religion tended to push aside philosophical involvement for the sake of an approach believed to be more biblical. We may hope that our study will lead us to some fresh thinking about this question. (2) As mentioned above, recent studies laid much weight upon the ideas of the nature of humanity held, or supposed to have been held, by the Hebrews and the Greeks (or, naturally, by other groups: but these two have been taken as the classic contrasting pair). Heavy theological weight was thus made to rest on judgments which had linguistic, sociological or anthropological character and over which theology itself scarcely had the necessary means for evaluation. (3) The first few chapters of Genesis form one of the biblical areas where there is an unusual degree of contact with myths and ideas known from peoples outside Israel. On the other hand, much modern biblical theology has striven to emphasize the separate identity and peculiar character of the Bible and to take these qualities as the very basis for claims about its authority within Christianity. Our study of the Garden of Eden will perhaps illuminate some part of this area also.

In addition, much scholarly research has been devoted in the last decades to the Jewish background of the New Testament texts, especially through the Dead Sea Scrolls but also on the basis of other documents. Much of this work, however, has been academic and fairly technical, and there has not been time for its effect to filter into general theological and popular discussion. This book also will remain outside the technical levels, but it will seek to let the influence of recent studies in Judaism have some effect on the more general understanding of immortality.

This book is not intended to be a thorough or systematic discussion of either immortality or resurrection. It follows out a certain group of biblical themes which I believe to be meaningfully related and linked to immortality. I have not attempted to engage with all the vast discussion in the scholarly literature or to cite every biblical passage that might be relevant; rather, I have preferred to restrict myself to a limited number of writers who seemed to form a creative circle. Most of my argument I worked out myself, without any large scholarly apparatus. In a few cases, after the lectures were delivered, I came upon writings that I had not known but that had taken similar paths: such was the valuable article of Ulrich Kellermann, and in particular it was a great pleasure to find opinions of similar tone expressed in Harold Bloom's comments in *The Book of J*, published in the same year in which the lectures were delivered. As commonly happens, the material tended to swell in volume after the lectures were delivered, and this will be particularly visible to the reader in ch. 2.

I owe particular thanks to Bristol University for honouring me with the invitation to deliver the lectures, and for its kindness and hospitality during my visit there. I am grateful especially to the Revd Simon Tugwell, OP, who assisted me by letting me see in advance a proof copy of his own earlier series of Read-Tuckwell Lectures, now published as *Human Immortality and the Redemption of Death* (Darton, Longman and Todd 1990); to Professor George Nickelsburg, who kindly made me a gift of the last available copy of his invaluable *Resurrection, Immortality and Eternal Life in Intertestamental Judaism* (Harvard 1972); and to Dr Robert Hanhart of Göttingen, who kindly obtained for me copies of the correspondence published in the *Kirchenblatt für die Reformierte Schweiz* of 1926, which is referred to on pp. 87 ff.

While in Bristol, during the period of the lectures, my wife and I greatly enjoyed the friendship and hospitality of the University, and the Departments of Theology and of Philosophy provided excellently for all the needs involved in the delivery of them. The five lectures were interspersed with seminars conducted on related topics by scholars of Bristol University, and these enlivened the series with variety and interest. I as lecturer gained and learned much from the five seminars conducted by Professor Ursula King on immortality in world religions, by Dr Lyndon Reynolds on aspects of mediaeval theology, by Dr David Milligan on the philosophy of belief in immortality, by Drs David Hopkins and T. A. Mason of the English Department on Milton's *Paradise Lost*, and by Professor C. J. F. Williams with a general response to the lecture series. Dr Margaret Davies took care of our everyday needs and was extremely kind and friendly. For us it was a wonderful occasion, and we wish all future prosperity and world-wide renown to the Read-Tuckwell Lectures.

In biblical passages, where the Hebrew numbering differs from the English, the Hebrew numbers are given first, the English in brackets afterwards.

Abbreviations

HTR	*Harvard Theological Review*
JBL	*Journal of Biblical Literature*
KBRS	*Kirchenblatt für die Reformierte Schweiz*
NEB	*New English Bible*
REB	*Revised English Bible*
SEÅ	*Svensk Exegetisk Årsbok*
SJT	*Scottish Journal of Theology*
THAT	*Theologisches Handwörterbuch zum Alten Testament*
VT	*Vetus Testamentum*
VTS	*Vetus Testamentum Supplements*

1

Adam and Eve, and the Chance of Immortality

The duty of the Read-Tuckwell Lecturer is to 'deliver a course of lectures upon Human Immortality and subjects thereto related', and this book is the result of my carrying out of that task. The first step, I think, is clearly that one should say something about the way in which these subjects have come to be viewed during the twentieth century. For in some ways our century has developed an understanding of immortality, and of eternal life, that is markedly different from that which was fairly normal for nearly two thousand years before. In what way have things altered?

The answer is essentially as follows. In the older religious tradition it was held as clear that the immortality of the human soul was central to religion. The human body was subject to sickness and death, but the soul was immortal and could not perish. Anyone who doubted the immortality of the soul was likely to be considered as a dangerous heretic, if not a total denier of religion. And even today it is likely that the average person, though himself or herself probably uncertain about immortality, still expects official religion to uphold and support it as an essential article of faith.

Those, however, who are *au fait* with modern theology, and especially with modern biblical studies, will be aware of a very different state of affairs. During the twentieth century it became almost a commonplace to maintain that the immortality of the soul was not part of biblical belief at all, and that the foundation of Christianity lay in the quite different doctrine of the resurrection of the body. Immortality was not only different from resurrection, not only a complementary viewpoint: it was actually opposed to it. Among the many voices arguing this case, we may note in particular that of the then influential Swiss exegete Oscar Cullmann, who published a short book arguing precisely this. His book, bearing the provocative title *Immortality of the Soul or Resurrection of the Dead?*, clearly suggested that the opposition between the two beliefs was close to absolute. Krister Stendahl, another leading New Testament scholar, though one whose general approach differed substantially from Cullmann's, was very interested in the topic and equally critical of the immortality of the soul. 'The whole world that comes to us, through the Bible, OT and NT,' he wrote, 'is not interested in the immortality of the soul.'[1] Opinions of this kind quickly became highly influential. Thus the important theologian Jürgen Moltmann, writing in the 1970s, tells us that 'resurrection of the dead excludes any idea of a "life after death" ... whether in the idea of the immortality of the soul or in the idea of the transmigration of souls'.[2] In this he represented a current of opinion that seems to have gone back to the beginnings of the 'dialectical theology' associated with the names of Barth, Brunner, Gogarten and Bultmann in the 1920s. In this tradition we find not merely doubts about the immortality of the soul but positive exultation in the denial of any such idea. Thus Wilhelm Vischer, writing in the 1930s about the book of Leviticus – not a place where one had customarily looked for ideas of immortality in any case – prizes Leviticus for its avoidance of them.[3] Expounding the concentration of Leviticus on *this life* in *this* land, Vischer notes with approval that there is 'not a word about a [life] beyond or about the immortality of the soul'. The most profound of other religions, such as the ancient Egyptian religion, he tells us, are religions of death.[4] By contrast the God of the Old Testament is a God of life. These arguments are typical of this current of thought. By the late twentieth century it seems to have become dominant. The only major recent theologian to question the validity of this trend, so far as I know, is Professor Maurice Wiles.[5]

This was undoubtedly a change of attitude in much theology. Oliver Quick, writing in the 1920s, was conscious of innovation: 'Our modern theology and philosophy have hardly perceived either the width of the gulf which separates a belief in resurrection from a belief in mere immortality, or by what providential guidance the Church was enabled from the first to stand firmly on the side of resurrection.'[6] For very many people, this was indeed something new. If one reads a work of representative theology like John Baillie's *And the Life Everlasting* (1933), written not long before the influence of dialectical theology began to be widely felt, one perceives a completely different viewpoint. The book is interested in the immortality of the soul, because the soul is the aspect of humanity which was especially adapted for communion with God. The question of the relation between that immortality and the resurrection of the body is scarcely mentioned. Not that Baillie has doubts about resurrection – far from it. But clearly it never enters his mind that there might be any opposition between the two. As he sees it, given that the resurrection of the body is an essential part of Christian belief, this means that we have to get ahead with discussing the many vital problems connected with the soul, its relation with God, and its immortality. It was this mode of approaching the question that was now swiftly to be swept aside.

Cullmann's little book remains the most effective marker of the change: partly because of its sheer concentration on the opposition between the two ideas, partly because of his status as a biblical scholar and his claim that the entire Bible was united in this respect, and partly because of his linkage of the opposition with the contrast between Hebrew and Greek ideas which just at that time was running at its high point of popularity. In biblical thought, according to this viewpoint, there is no separation of body and soul: body and soul are a unity, and it is the resurrection of the body that is central to faith. The idea of an immortal soul comes not out of the Hebrew heritage of concepts upon which Christianity was built, but out of Platonic philosophy, and it is a severe distortion of the New Testament if this foreign idea is read into its teaching. The immortality of the soul, then, in so far as it has been part of the Christian tradition, has entered into it through a mistaken infiltration of Greek philosophical ideas into the quite different idea-world of the Bible.

Now whether these arguments are justified or not, it is too early for us to say. I state them here simply as a necessary background to the discussions that are to follow. We may begin by recognizing this: that while some traditions of theology, and especially of philosophical theology, have continued to be very interested in the theme of immortality, others, and especially important trends in the use of the Bible within theology, have tended to become hostile to the entire idea of it and to disregard it as an element in biblical thought. Since I am a biblical scholar, and intend to approach the whole subject from the biblical side, I have to begin by explaining all this. And to these general questions we shall return later. But our first avenue of approach, as the title of this chapter indicates, will be through one particular biblical narrative of great power and centrality, the story of Adam and Eve and their expulsion from the Garden of Eden.[7]

So let me put forward at once my basic thesis about that story. My argument is that, taken in itself and for itself, this narrative is not, as it has commonly been understood in our tradition, basically a story of the origins of sin and evil, still less a depiction of absolute evil or total depravity: it is a story of how human immortality was almost gained, but in fact was lost. This was, I need hardly remind you, the reason, and the only reason, why Adam and Eve were expelled from the Garden of Eden: not because they were unworthy to stay there, or because they were hopelessly alienated from God, but because, if they stayed there, they would soon gain access to the tree of life, and eat of its fruit, and gain immortality: they would 'live for ever' (Genesis 3.22). Immortality was what they had practically achieved.

Now in order to understand this we have first of all to look at the interpretation that has been more dominant and familiar, according to which the story tells of the 'Fall of Man', a sudden, drastic and catastrophic change by which the human relationship with God was ruined. This understanding derives essentially from St Paul. To him the total and unqualified gift of salvation through Jesus Christ was the reversed image of the equally total and unqualified disaster brought about by Adam and through Adam transmitted to the entire human race. By one man sin entered the world, and through sin death, and thus death pervaded the whole human race, inasmuch as all men have sinned (Romans 5.12). Human mortality, then, is a consequence of sin, and this leads by a straight line to Christ's death for sin, and to his overcoming sin through overcoming death. 'Original sin' is a familiar term that has grown out of all this: sin goes back to the remotest beginnings of humanity, and is inherited by us all. There were never any humans who were not sinners, apart from Adam and Eve, and these only for a very short time. It is easy to see how Paul, wishing to make clear the completeness and finality of Christ's victory over sin, looked to the story of Adam and found in it the typology that he needed.

It is important to perceive that this analogy is very much Paul's own property. So widely established did it become in Christian thought and tradition that one does not easily become aware of its narrow basis within the New Testament itself. The typology of Adam and Christ is absent from the teaching of Jesus, from the Gospels in general, from the other Johannine literature, from Hebrews, Peter and James, from everything. Jesus himself, though he noted some features of the early Genesis story in other respects, shows no interest in Adam or Eve as the persons who brought sin and death into the world. Apart from Paul, Adam is mentioned little in the entire New Testament and only incidentally. And even in Paul the Adam–Christ typology is not so very widespread: explicit mention of it is confined to Romans 5, I Corinthians 15 and I Timothy. Clearly, the emphasis on the sin of Eve and Adam as the means by which death came into the world was not considered a universal necessity in New Testament Christianity: whole books were written which took no notice of it. It is a peculiarity of St Paul, and it is very likely that the thought originated with him; or, to be more precise, that its use as an important element *within Christianity* originated with him.[8] For, as we shall see, Paul had predecessors in Jewish literature who prepared the way for his ideas. Powerful, then, as the Adam–Christ comparison was to prove, it was not an essential structure of the earliest Christian faith but was a part of the typology which one particular person or tradition found helpful for the expressing of an understanding of Christ.

Such a typology was in every way creative, so long as it was seen to be no more than that. But the typological use of the figure of Adam was not the same as the detailed explication of every aspect of the story. There were, as we shall see, many details and aspects of the story which did not easily fit in with the Pauline exploitation of it. Of these the most obvious and important lay in the place of *death*. As most traditional Christianity has read the story, the humans were sinless and free from all threat or reality of death, and then through their one disobedience they instantly came under the total dominion of sin and became subject to the power of death. Read for itself, in its ancient Hebrew cultural context, the story may well say the opposite. The story nowhere says that Adam, before his disobedience, was immortal, was never going to die. The natural cultural assumption is the opposite: to grow old and die with dignity, surrounded by one's children and family, was a good and proper thing, to which Adam no doubt looked forward as anyone should. To die was human; only gods – generally speaking – lived for ever.[9] Conversely, however, the problem that Adam's disobedience created – a problem for God himself above all – was not that he brought death into the world, but that he brought near to himself the distant possibility of immortality. This alone is the reason why he and his wife have to be expelled from the Garden of Eden.

But before we go farther with the story in itself let me mention several qualifications which should be set against the common tradition under which the story is seen as the tale of 'the Fall', the original sin through which all sin and death in human life and history came about. First, it is not without importance that the term 'sin' is not used anywhere in the story – its first appearance in the Bible is in a somewhat obscure phrase in the story of Cain, Genesis 4.7 – nor do we find any of the terms usually understood as 'evil', 'rebellion', 'transgression' or 'guilt'. The story proceeds on the level of practical actions: 'if you eat this, in that day you will die', 'God made them coats of skins', and so on, and the express verbalization of any of this as 'sin' or 'origin of evil' or the like is lacking.

Secondly, within the Hebrew Bible itself the story of Adam and Eve is nowhere cited as the explanation for sin and evil in the world. This reference, which to us seems so natural, simply does not occur. This is striking, because the Hebrews were quite capable of referring back to the beginnings of things: thus 'Your first father sinned' (Isaiah 43.27), though the reference there is not to Adam, but to Jacob or some other pioneer of the people of Israel.[10] The Old Testament is certainly deeply conscious of the actuality and pervasiveness of sin and evil. But nowhere in all the books of the Hebrew canon is the existence or the profundity of evil accounted for on the grounds that Adam's disobedience originated it or made it inevitable. It is not clear that the Old Testament is interested in knowing or finding one universal cause or origin of evil.

Indeed, far from taking the universal sinfulness of humanity as an obvious and ineluctable fact, the Old Testament seems to assume the possibility of *avoiding* sin. 'In sin did my mother conceive me' (Psalm 51.5, Hebrew 51.7) was always a favourite text for the orthodox Christian point of view. But the first striking thing about it is how isolated it is, how few passages there are that even appear to say the same sort of thing. No other passage, perhaps, exists which could be quoted to indicate something like a universal environment of evil in which all persons were born.

That such an opinion existed need not surprise us. Ideas that all persons sinned were widespread: the idea is not at all original in the Old Testament, but is part also of the environing religious atmosphere. A Mesopotamian worshipper, addressing Marduk, says:

Who has not sinned, who has not committed offence?
Who can discern the way of the gods?[11]

The idea that all humans were sinful certainly existed, and Psalm 51 is a powerfully concentrated example of it. But it cannot be taken as a carefully thought out and universal theological principle. For within the same culture, and in particular within the Old Testament, it lay alongside other ideas which pointed in a very different direction. It is only because people read the Psalms through the glass of a later perspective that they see them as poems deeply imbued with the sense of sin and the need for atonement. Sometimes indeed this is so, but often the reverse is to be found. The striking thing about them, in many places, is the poet's insistence that he (or the worshipper for whom he speaks) is free from blame and guilt. There are sinners and evildoers everywhere, of course, but the poet himself is not one of them, nor is the worshipper who uses the poem.

The Lord rewarded me according to my righteousness ... for I have kept the ways of the Lord and have not wickedly departed from my God ... I was blameless before him, and I kept myself from guilt' (Psalm 18.20–23).

Again, in Psalm 26 the poet repeatedly claims to have 'walked in his integrity' (vv. 1, 11): there are sinners all around him, but he is not one of them. Paul in Romans 3.9–18 argues that 'all men, both Jews and Greeks, are under the power of sin' (RSV), and quotes a series of passages from the Psalms and Isaiah, but a glance at the original forms of these passages makes it clear that the sin referred to was not universal and that there were two groups in view, the sinful on the one hand and the righteous on the other. Thus, for example, Paul quotes Psalm 14.1–2 to the effect that

None is righteous, no, not one,
no one understands, no one seeks for God ...

but the Hebrew poem makes it very probable that these remarks apply to the 'fools' who appear in the first few words, and not to the totality of mankind. We do not have to look farther than v. 5 of the same poem to see a reference to God as being

with the generation of the righteous

– this 'generation' is clearly a group that stands in contrast with the 'fools' who say that there is no God and who form the object of the Psalmist's disapproval. Paul obtained his demonstration that sin was universal, that 'all, both Jews and Greeks' are 'under the power of sin', only through taking the phrase 'none is righteous, no, not one' out of the context of the poem in which it stands.

Again, Psalm 17 shows the singer inviting God to try him, confident that God will find no wickedness in him. 'I have avoided the ways of the violent,' he says, 'My steps have held fast to thy paths, my feet have not slipped.' The point can be repeated many times.

The book of Job shows the same assumptions. The whole atmosphere of the Hebrew Bible works against the idea that sin and evil were taken, as a matter of theological principle, as something that belonged of necessity to all human life. Violent outbreaks of evil, yes, but a steady unchanging subjection to it, no.

It is not surprising therefore that the main Jewish tradition, as we know it since the Middle Ages, has refused to accept any sort of doctrine of original sin. It does not follow, however, that no such ideas ever existed in earlier stages of Jewish tradition, and we shall see that they did indeed exist. Interpretations that look quite like an idea of universal original sin did arise and can be found in Jewish traditions from between the Old and the New Testament periods. It is, however, possible to read the canonical Old Testament, taken in itself, in a way that sees nothing like original sin as an important element in its purport. Of course the Old Testament has page after page of evil deeds, and disasters that follow failure to pursue the will of God are practically normal within it. But these are *actual* evils, not evils that are necessitated by a given inheritance, a propensity towards evil which humans cannot of themselves overcome. So at least it can be read.

But still more central to our theme is the matter of *death*. The centrality of death is emphasized from the beginning. God said to the man, before the woman came on the scene, 'You must not eat of the fruit of this tree; for on the day that you eat of it you will certainly die' (NEB). Later the snake says to Eve: 'You will certainly not die. For God knows that when you eat of the fruit your eyes will be opened, and you will be like God, knowing good and evil.' They ate of the fruit; their eyes were opened, they found that they knew good and evil; and they did *not* die. The serpent was the one who was right in such matters. They did not die. Indeed, the punishment brought upon the man does include the *mention* of death: because of man the ground is cursed, and he will suffer toil and frustration all his life. In the sweat of his face he will eat food, until he returns to the ground, for from the ground he was taken, and to dust he will return. Yes, indeed, but this is not death 'in the day that' they disobeyed, it is not death in itself that is God's response to the disobedience: rather, the punishment lies in the area of *work*. Work itself is not terrible; Adam was placed in the garden in order 'to work it and to keep it', no doubt an easy job under ideal conditions, picking fruit from the trees as he worked. His punishment is that *the ground* is put under a curse: it will produce weeds, and Adam will have to live off grasses, which will be intermixed with these weeds, making for endless toil and lack of success. It is pain and failure in work, toil and frustration in toil, and the final frustration is death, the final proof, far off in the future, that all his work will get him nowhere. On the contrary, his death will mean his own returning to that same refractory soil which has made his life so bitter. His death is not the punishment, but is only the mode in which the final stage of the punishment works out. He was going to die anyway, but *this* formulation of his death emphasized his failure to overcome the soil and his own belonging to it.

The death to which Adam will finally fall victim, then, his 'returning to the dust' (for the term 'death' is significantly not used at this point), is not in itself a punishment, as many scholars have long seen. That this is so is manifest if we consider the reactions of God to the snake and to the woman. Take the woman first: she was certainly as much involved in the disobedience as the man had been, she was the one who had talked with the snake and been persuaded by him, she had done the thinking about the matter in so far as any thinking was done at all, she had taken the forbidden fruit, and she was the one who had given it to her husband. So if anyone was to be punished with death she was a likely candidate. But *her* sentence of punishment says nothing about death. She is certainly to have a bad time: increased pain in childbirth, desire for her husband, and some sort of domination by him – all of them no doubt unpleasant effects, but coming far short of being put to death. And so also the snake, who certainly had a considerable responsibility for the whole affair: he was to crawl on his belly all the days of his life, instead of walking about upright as he must have done beforehand, and to eat dust; and enmity between him and humanity was to be eternal. But once again this falls far short of immediate execution. The seed of the man might crush the serpent's head, but the serpent would get some of his own back, snapping at the human's heel. Snakes might well enjoy this way of life. There was no death here. And if this was true of the snake and the woman, it was true of the man: his returning to dust is part of the picture of his bitter agricultural life; in the end, after all his struggle with the unrewarding land, he would himself be swallowed up in it and become part of it. The 'returning to the dust' is mentioned only of Adam, precisely because, of the three, he alone was the agriculturalist who would have to struggle with the stones and weeds of the land.

And let us add at once another point, to forestall a possible objection: it is not legitimate to argue that, though Adam and Eve did not actually die immediately after their disobedience, they were nevertheless 'condemned to death' and thus God's original warning that they would die in the day that they ate of the tree was indeed fulfilled. That is an evasion of the text and its evidence. What God in the beginning said was a warning, a warning of a kind amply paralleled elsewhere in the Bible: 'if you do this, you will certainly die' – clearly, as an instant punishment. This kind of warning is a warning for *mortals*, and is well evidenced elsewhere in the Bible. Mortals know that they will die, eventually; what they do not want is to die *now*, or in the near future. Such warnings make sense only if the punishment for disobedience is speedy. God's warning can only mean death soon after – not necessarily within twenty-four hours in a totally literal sense, but certainly a prompt or immediate reaction. The phrase is parallel with other similar warnings in the Bible: for instance, Solomon's warning to Shimei in I Kings 2.37,42, literally 'in the day that you go out and cross the Kidron valley, you surely know that you will certainly die'. Perhaps he was not put to death on that very day; it may have been a day or two before Solomon knew of the incident. But the execution was not delayed. The warning would have been of no value if it had meant 'well, at some time in the next twenty or thirty years you will die'. None of us will be deterred from evil-doing if we are told that 'if you do this, you will be a dead man (or woman) a hundred years from now'. Since Adam lived on for a long time, to the age of 930 years according to Gen. 5.4–5, to say that his death at the end of this very long life, the fourth longest in the Bible, fulfilled the warning that he would die 'in the day' of his eating the forbidden fruit cannot be taken seriously. The warning was one of speedy punishment. But this necessarily means that the threatened punishment was not carried out. Of course Adam died; but he would have died anyway. By one account he would die, miserably absorbed in the refractory soil with which he had struggled all his life; by another, he died at the age of 930, full of years, probably a more positive picture. In either case he was to die from the beginning. Possibly the manner of his dying was the consequence of his disobedience, but his dying in itself was not.

And this is not the only circumstance that indicates that the origin of sin and evil is not the only, or the main, interest of our story. Another is the absence from the text of the atmosphere of guilt and tragedy. Not only, as we saw above, is the vocabulary of guilt and revolt lacking, but the atmosphere of catastrophe is also absent. In contrast to Cain, who lamented that his guilt or punishment was too heavy for him to bear (Genesis 4.14), no such sentiment is voiced by either Adam or Eve. God is indeed angry, for the humans have disobeyed his command, and they are naturally frightened of him and hide from him among the trees. But there is no breakdown of relationship between God and them. They continue to talk on normal, if irritated, terms. It simply is not true that relations between God and the human pair have broken down.

Now the mention of the nakedness of the man and woman has often been understood as if this expressed the sense of guilt. But this again is not necessarily so, and the reading of the text in this way may well be the effect of centuries of Christian experience which has perceived nakedness in that sense. Their perception of nakedness follows directly from their acquisition of *knowledge*. Knowledge brings self-consciousness. Before the disobedience they were naked and had no embarrassment, because they lacked this sort of knowledge. In this respect they were like the animal world. Once they have knowledge, they cannot continue unclothed: they hide from God because they are naked, and he is quick to make them garments to cover them, for the normal human reality demands clothes. The whole matter of the nakedness and clothing does not require interpretation in terms of a serious feeling of guilt at all. But this aspect will be taken up again in a later chapter and I will say no more about it at this point.

And this is confirmed by another feature, namely the somewhat unserious aspect of the whole thing. This was noticed long ago, and is well exemplified by those critics who complained 'What a fuss about a mere apple!' For the God who places upon humanity the one condition, that they should not eat from a particular tree, is a God who is not insisting upon any very central ethical principle. Eating that fruit is not in the same category of offences as murder, which was Cain's offence, or filling the earth with violence, which was that of the generation before the Flood. Why was it wrong to eat that fruit? In fact, we are left to surmise, because God wants to keep to himself the knowledge of good and evil: he does not want anyone else to have it, and still more does he want to keep to himself the reality of eternal life. The sheer irrationality of the command, not to eat of the tree, and of the threat to deprive of life if it was eaten, has had great effect on the history of understanding: for it has been read as if to mean that the slightest deviation from the slightest divine command, however devoid of perceptible ethical basis that command might be, was and must be a totally catastrophic sin which would estrange from God not only the immediate offender but also all future descendants and indeed all future humanity. None of this, however, is involved in the actual story. It is God who is placed in a rather ambiguous light. He has made an ethically arbitrary prohibition, and backed it up with a threat to kill which, in the event, he does nothing to carry out. He is of course angry, and the man and woman are frightened. But, after issuing announcements of humiliations, limitations and frustrations to which they will be liable, he goes on to care for them and provide the necessary clothes, before proceeding to expel them from the Garden of Eden. The story has a mildly ironic and comic character rather than one of unrelieved tragedy and catastrophe. Bloom speaks of the 'ironic joys' of the writer J,[12] and let me quote a longer statement by Sean McEvenue:[13]

> Distance between the author and the reader is maintained by a witty style. The reader is never invited to identify, or sympathize, with anyone in the story. The reader feels the cleverness of the author, but does not feel any weighty message from the author. The story is about sin and punishment, but no narrative slant or rhetorical device is used to make the reader feel fear, or guilt, or sorrow. The treatment of sin or evil is not marked by theological or philosophical depth; rather its most memorable passages, the treatment of temptation in 3:1–6, and of guilt in 3:7–13, are characterized by a dramatic subtlety which one associates more easily with comedy than with tragedy.

This is confirmed when we consider another factor, namely the *motivation* of the woman and the man – particularly the woman, since the text gives no indications about the man's motivation. Here again traditional theology had a clear and simple explanation, which however is ill-supported by the narrative itself. Especially in the West, and since St Augustine, and increasingly so, if anything, in Protestant theology, if that were possible, the explanation given is *pride*, *superbia*, the will to be more than human, the desire to transcend the limitations of humanity and be like God, above and outside the world. This may well be an important imaginative insight and might, indeed, have much truth in it as a theology of human nature. But it is ill based in the story of Adam and Eve. The story gives no evidence for the idea that they longed to be as gods. It is true that the snake, in talking to Eve, points out that God knows that as soon as they eat the fruit their eyes will be opened and they will be like gods, knowing good and evil. And, like other things that the snake said, this was no doubt true enough. It was, from the snake's point of view, a statement about God's motivation: God did not want humans to be as gods, knowing good and evil. But this does not mean that the desire to be divine was the motivation of the humans. Even though the woman heard this remark, it is not clear that she was much attracted by the prospect of being like a god. She saw that the tree was (*a*) good for food, (*b*) nice to look at, and (*c*) excellent for giving wisdom (this last point slightly uncertain, since other meanings are possible, but let it pass for the moment). She took of the fruit and gave some to her husband. There is nothing here about lusting for superhuman status and power, nothing about taking over from God the task of governing the world. The woman's motives were distinctly within the normal limits and passions of humanity. It was the dietetic aspect (the nutritious value of the food), the aesthetic aspect (its good appearance), and the educational aspect (its ability to give wisdom) that attracted her. There is nothing here of a rebellion against God, nothing of a titanic will to take over the status of the divine. The motivation is sketched with a noticeably light touch. Even the knowing of good and evil, which the serpent has actually mentioned as the expected result of eating the fruit, is not included as an attraction which actually moved the woman's mind. The nearest one comes to it is her noting that the fruit is good to make one wise. If the woman is to be censured, it is more for concentrating on short-term attractions like food-value, appearance and educative assistance, when she might better have been thinking about long-term matters like the knowledge of good and evil. In the result, the motivation is less that of aspiration to divine status, still less that of rebellion against God; it is more a mixture of physical attraction, curiosity and insouciance or inadvertence. It is as if you have in the house a large red switch on the wall with a notice saying 'This switch must on no account ever be touched', and then one day there comes an imposing official in uniform with gilded cap, and you ask about the switch and he says, 'Well, of course they *say* you mustn't touch it, but they are just saying that: of course you can throw the switch and no harm will be done, indeed your electricity will probably run all the better if you do.' So of course you throw the switch and Bang! up goes the house in smoke.

Thus, to sum up, nowhere in the entire story do either of the human pair show any longing to escape the limitations of their humanity. The tones of rebellion against God are lacking. Thus once again we see that the origin of sin and evil is not the major theme, and, this being so, we are left with the two themes which are made explicit in the story itself, namely knowledge and immortality. And this brings us to another aspect of the same kind. The person who comes out of this story with a slightly shaky moral record is, of course, God. Why does he want to keep eternal life for himself and not let them share it? Even more seriously, why does he not want them to have knowledge of good and evil? What is wrong with this knowledge, that they should not possess it? And if we look at the traditional interpretation again for a moment: if Adam and Eve were, before their disobedience, completely free from death, why then under these circumstances were they to live eternally without any knowledge of good and evil, to be a sort of perfect human toy, living in a paradise where no moral sense was ever required or to be required? If God created these supreme earthly beings, who from the start had eternal life or immortality, and yet these remarkable persons were to be without moral sense or any need for any moral sense, is this much to the credit of the deity? No, everything points in the other direction: Adam and Eve were mortals, as human beings normally were, but through disobedience or mischance, perhaps of a relatively minor nature, they came near to the achieving of eternal life. The importance of this for our subject is great, for it means that in the structure of biblical ideas immortality does not come in at the margin, at the latest point, or through the intrusion of Greek philosophy. It is present, at least as an idea, at the earliest stages, and is a force that thereby has an effect on much of the thought of later times.

That this sort of interest in immortality, in the sense here implied, is ancient in the Old Testament seems highly probable. It is something that goes back to, and links up with, a wider and older mythological background, only part of which remains alive in Israel of biblical times. In this world it is obvious that humans are mortal; only gods are immortal. But there are occasional exceptions, or at least such exceptions are not unthinkable. The Bible has two or three of this class. The two clear and certain cases are Enoch and Elijah. Of Enoch, who was the seventh in succession after Adam, we are told that he 'walked with God', and 'he was not', was no longer there, for God took him (Genesis 5.22,24). He was then 365 years old, and unlike the other patriarchs no death of him is recorded: he did not die. In New Testament times this was still remembered: 'By faith Enoch was translated that he should not see death' (Hebrews 11.5). And similarly Elijah in II Kings 2: a heavenly chariot took him up, and the interesting thing is that though this was a highly exceptional event, it was at least partly known of beforehand by the prophetic group, who spoke of it without surprise.

A third case is not of quite the same category: that of Moses, of whom it is reported that his grave was unknown, and the text says, enigmatically, 'and he buried him' (Deuteronomy 34.6), leaving open the question of who it was that buried him: was this another supernatural removal? Hardly so in the OT itself, but later certainly, notably in the *Assumption of Moses*, where Moses tells us:

> For from my death, that is, my assumption, until his advent [that of the Heavenly One] there shall be two hundred and fifty times (*Assumption of Moses* 10.12).[14]

This might seem strange, but the Tannaitic midrash Sifre has the explicit statement: 'There are those who say that Moses did not die but stands and serves on high'.[15] The idea was known to Josephus and to Philo. The former wrote that Moses, suddenly covered by cloud, 'disappeared' in a ravine; to avoid leaving the impression that he had 'gone back to the deity', πρὸς τὸ θεῖον ἀναχωρῆσαι (the same expression used earlier of Enoch), he had written of himself that he had died. The same possibility had been considered when he stayed too long on Mount Sinai and it was thought that he would not come down again.[16] Philo, pointing out that Moses' death is not reported with the usual formulas (he must mean 'breathed his last' [Hebrew גוע] or 'be gathered to one's fathers'), thinks it fitting that it should be described with the word μετανίσταται, 'is translated', closely similar to the μετετέθη used of Enoch in Hebrews 11.5.[17] As Goldin says in discussing this, 'It is not a serious problem that it is distinctly stated [that] Moses died. On a number of occasions the Midrash does not hesitate to go its independent way, ignoring the biblical plain-spoken statement.' And he goes on to give reasons why, from within the thinking of Judaism, these rather unexpected thoughts about the death of Moses arose. The importance of the Assumption of Moses is, of course, indicated by the quotation of it, or at least a use of a related legend, in the New Testament (Jude 9). Yet more, the selection of Moses to be with Elijah on the Mount of Transfiguration may have been influenced by the understanding that both of these were still living from ancient times.

What I am saying is that Adam and Eve came near to belonging to this same class. But there was a difference, in that the others gained eternal life through their exceptional service to God, while Adam and Eve, if they came close to gaining it at all, did so (*a*) not through exceptional service but through a disobedience, and (*b*) more or less accidentally or incidentally, in that through another matter altogether they gained the ability to get to the tree of life. In the other cases it depends on something different: that is, the possibility of eternal life, extremely rarely granted, depends on the power and will of God to grant that life.

Now this argument has emphasized the centrality of immortality in the story of Adam and Eve, and in so doing it has de-emphasized the aspects which since St Paul have been most heavily stressed, namely the ideas of sin and the origin of evil, especially the idea that sin brought death into the world and placed all humans under the power of sin and death: I have argued that in Genesis itself the man and woman were mortal from the beginning, and their disobedience created a problem because it brought them to the possibility of eternal life. But the Pauline understanding also has its origins and background in writings earlier than Paul. Much the clearest antecedent of Paul is the Wisdom of Solomon: it most clearly stated the parameters that were to be regulative for St Paul:

> God created man for incorruption (ἐπ' ἀφθαρσίᾳ)[18]
> and made him in the image of his own eternity;

but through the devil's envy death entered the world,
and those who belong to his party (μερίς) experience it.
(Wisdom 2.23)

All the foundations of the Pauline picture of Adam are here. Humanity was created immortal or at least to be immortal. God being immortal, the image of God in humanity was a share in that immortality. The snake of Genesis was the devil. It was through his intervention (and the consequent sin, which is not specified) that death entered the world.

We do not know with certainty whether Paul had read the Wisdom of Solomon; but it does not matter much, for if he had not read it, that only means that he followed that same tradition of understanding which it was the first writing now extant to express. Whether he counted it as 'Holy Scripture' or not – the idea that this term itself was already clearly defined may well be anachronistic[19] – he accepted it, or the tradition which it represents, as a valid theological guide. And, given that this understanding was to be accepted – and, indeed, it is a simple and powerful overall statement – it was easy for the many features in the Genesis text which pointed in another direction to be overlooked.

This does not end the importance of the Wisdom of Solomon for our purpose. For, first of all, of the books now generally deemed (in Protestantism) to be 'Apocrypha', it is undoubtedly among the most Hellenistic in style and thought. This is highly significant for the evaluation of Paul's interpretation. Secondly, of all the books it is the one which has the greatest interest in precisely our subject: immortality. Not only does it make clear that man was immortal – or at the least was destined for immortality – before the first disobedience, but it dwells on immortality again and again as a guiding factor, an ethical constraint, and a supreme goal. Wisdom links together, as no other work does, the two aspects of our theme: the Garden of Eden and the hope of immortality. Thirdly, among all biblical or near-biblical books, it devotes especial attention to the theme of idolatry, and traces a clear linkage between idolatry and all later immorality, in a style which is very closely followed by Paul in the letter to the Romans. Fourthly, this attack on idolatry is built upon the foundation of a natural theology, again closely similar to the argument of Paul in Romans.[20]

The Wisdom of Solomon is not the only book of that late

Hellenistic epoch to be relevant. Equally important is IV Ezra, for it dwells most upon the continuing dreadful inheritance of Adam's sin upon all his later descendants. It is here, among the biblical and near-biblical books, that we come closest to the idea of universal and 'original' sin:

> Better had it been that the earth had not produced Adam, or else, having produced him, [for thee] to have restrained him from sinning ... O Adam, what have you done! For though it was you that sinned, the fall was not yours alone, but ours also who are your descendants! (IV Ezra 7. 116–118)

or likewise:

> The first Adam, loaded down with an evil heart, transgressed and was overcome; and likewise all who were born of him. (IV Ezra 3.21)

Thus, to sum up in advance an important part of our argument, it is in certain later strata of the Old Testament, including the books that are outside the present Hebrew canon, that the real grounds for the Pauline understanding of Adam and Eve are to be found. In Genesis, on the other hand, it is poorly supported. Far from it being the case that Paul's thinking is deeply rooted in the thought world of ancient Israel, it is much more precisely formed by the *interpretation* of these ancient texts which took place in Hellenistic times and in a different intellectual atmosphere.

The reader may think that these are somewhat revolutionary, isolated and idiosyncratic opinions, but this would be a mistake. On the contrary, they represent a substantial consensus position among central Old Testament scholars. Much of it is implied, if not exactly stated, in the detailed commentary of Westermann on Genesis, and I quote the words of Professor M.A. Knibb in a recent competent survey:

> There is no suggestion in the narratives of the creation and fall, nor indeed in the Old Testament as a whole, that man was created immortal and lost his immortality as a result of disobedience. In Genesis 2.17 death is certainly prescribed as the penalty for eating the tree of the knowledge of good and evil, but there is no hint that man had originally possessed immortality.[21]

Thus the position I argue here has a wide basis in accepted exegetical thinking, especially on the negative side, that is, the refusal of the idea that man was created immortal. The only difference, if any, is that, taking these established exegetical positions as basic, I go on to lay a stronger emphasis than has been usually done on the theme of immortality as a chance that was lost.

I am saying, then, that the thought of a possible, at least a conceivable, immortality for humans is a central aspect of these parts of the book of Genesis. In saying this I am on the one hand entering into the mythological world out of which the story has grown, and I shall hope to say something more about that mythological world in the course of this discussion. But on the other hand I am not placing the entire stress on the 'original' or reconstructed prehistory of the story, for I am saying that the story in its final, canonical form itself makes clear that it conveys the same emphasis. The notion of a human immortality, that might conceivably have been gained but was in fact missed, stands both in the pre-biblical mythology and in the Genesis text, where it takes a most significant place as the start of the canonical narrative of human existence. Moreover, I believe that this insight is of high importance for the religious history within Israel. In particular, our story does not speak of 'life after death', nor about the 'immortality of the soul'. The 'living for ever' which Adam and Eve would have acquired had they stayed in the garden of Eden is a permanent continuance of human life.

It is striking how difficult it has been for scholars to give an adequate account of the history, within Israel, of beliefs in resurrection and immortality. One reason for this, I suggest, is that they have too often been influenced by the conception that what they are looking for is something *after death*. And life after death does certainly appear within the tradition. But the starting point seems to me to lie rather in the immortality that is normal for the gods, an immortality where there is no death anyhow, an eternal life that is normal for the gods but might, in very exceptional circumstances, have been granted to humans. To these historical aspects we shall return.

If this is the root from which later ideas have grown, it differs in many ways from these later ideas, and most obviously in the respect I have just mentioned: namely, that immortality here, or eternal life, in the first place does not mean life *after* death, but the continuance of life without death. And secondly, I have not yet said anything about the 'immortality of the soul', for thus far questions about soul and body have hardly come into the picture. Before we can say more about these aspects, we have to look at the thoughts of the Hebrews about death, and that will be the starting point of our second chapter.

2

The Naturalness of Death, and the Path to Immortality

I have argued, then, that the central theme of the story of the Garden of Eden, along with the theme of knowledge of good and evil, is the theme of immortality. It is because Adam and Eve, if they had been left in the garden, would have eaten of the tree of life and 'lived for ever' that they were expelled. The chance of immortality was lost. This does not mean that the story is written out of longing for immortality: it is not. On the contrary, its purport is the acceptance of mortality. Mortality was characteristic of human life from the beginning, and the very rare possible exceptions do not alter that fact. The same was true in the world of Homer: there were very exceptional cases where a mortal was granted to live for ever. Odysseus, if he had consented to wed a nymph or minor goddess, would have achieved this. But this did not make any difference: all humans were well aware that death awaited them and could not be turned aside. Nevertheless, in the Hebrew situation that momentary glimmering of immortality may have been of some importance in the later development of belief.

Thus we return to the more general question of the 'naturalness' of death for humanity. If I am right in what has been said, mortality was the 'natural' lot of humanity from the beginning. Adam and Eve were not immortal beings who by sin fell into a position where they must die; they were mortal beings who had a remote and momentary chance of eternal life but gained this chance only through an act of their own fault, and who because of that same act were deprived of that same chance. But by formulating things in this way, we have to consider how they relate to the entire question of the biblical view of life and death, the 'naturalness' and the potentialities of each of them.

Here we can resume with another quotation from Cullmann's

famous and influential booklet. He insists on the *unnatural* character of death. I quote a long passage at a key point in his argument (p. 28):

> Yet the contrast between the Greek idea of the immortality of the soul and the Christian belief in the resurrection is still deeper. The belief in the resurrection presupposes the Jewish connexion between death and *sin*. Death is not something natural, willed by God, as in the thought of the Greek philosophers; it is rather something unnatural, abnormal, opposed to God. The Genesis narrative teaches us that it came into the world only by the sin of man. Death is a curse, and the whole creation has become involved in the curse ... Death can be conquered only to the extent that sin is removed. For 'death is the wages of sin'. It is not only the Genesis narrative which speaks thus. Paul says the same thing (Romans 6.23), and this is the view of death held by the whole of primitive Christianity ... To be sure, God can make use of death ... Nevertheless, death *as such* is the enemy of God.

Now there are serious weaknesses in this kind of argument. For one thing, if the 'Jewish connexion between death and sin' is so important for the Christian idea of resurrection, why is it that the Old Testament itself is so limited in its own affirmations of resurrection?[1] Scholars searching for the development of ideas in this area have a hard time providing good material. It is commonly accepted that the earliest definite expression lies in Daniel 12, one of the latest texts of the Old Testament, and even it is somewhat vague: 'many', it says, of those who 'sleep' in the dust of the earth 'shall awake', some to everlasting life, and some to shame and everlasting contempt (Daniel 12.2). This is far from a ringing affirmation of bodily resurrection as against other depictions of survival or continuance after death. It says nothing that opposes or excludes such alternative ideas as the immortality of the soul. It says nothing about the rest of humanity, apart from the 'many' who thus 'sleep' in the dust of the earth – who could, after all, be a special group. On the whole, one supposes, the Cullmannian position would have to be rephrased thus: it is the *negativity* of the Old Testament towards all ideas of after-life that provides the necessary presuppositions for the meaningfulness of resurrection. And that is not necessarily a mistaken argument. But at best it remains a narrow and negative basis for the linking of resurrection with the Hebrew Bible.

Secondly, in fact, as has been amply shown by a variety of writers, when we make a thorough investigation of Jewish ideas of life after death through the period of Daniel and down to the beginnings of Christianity, the evidence shows that a wide variety of opinions existed: resurrection of the body, immortality of the soul, an 'eternal life' marked by quality within the present life, and others. Each of these could be conditional and subject to variations and combinations. Resurrection might be for a few; and it might be only for the righteous, no counterpart being granted to evildoers. Immortality might be a quality of humanity from the beginning, but it might be something that had to be *gained* by proper conduct. It, too, might be limited to a particular group. 'Eternal life' in the sense of quality within this life might lead on to life after death or to resurrection, or it might simply ignore the question of death. Right down to the beginnings of Christianity, there was, it seems, no fixed or definite view about these matters. Jesus' disciples themselves appear to have had no expectation of a resurrection of their Master until it took place. In particular, as I shall show, thoughts of the immortality of the soul were powerful within late Judaism. Even where 'the Jewish connexion between death and sin' was fully recognized, there was no idea that this necessarily implied the option of bodily resurrection and excluded the option of immortality of the soul. Cullmann's bypassing, within this argument, of the intertestamental material – which, of course, he must have known quite well – lays him open to devastating criticism: 'By excluding any discussion of specific Jewish texts, Cullmann has approached the New Testament presupposing a unitary Jewish view which is a pure fiction.'[2] In spite of his distinction as a New Testament scholar, in this matter and in respect of the Hebrew and Jewish connections he seems simply to have assumed the validity of the older dogmatic traditions and ignored the evidence of texts.[3]

Thirdly, as the first chapter has made clear, Cullmann was naive in saying that Genesis 'teaches us' all this and not only Genesis: 'Paul says the same thing'. It is purely because Cullmann takes Paul's use of Genesis as normative that he thinks that any of this is 'taught' by Genesis. As I have shown, Paul's own handling of the theme of Adam and Eve fits poorly with Genesis and can be justified only if we move to the ground, not of Genesis in itself, but of trends in the use of Genesis which are known to us first from the Hellenistic era. Thus, disastrously for Cullmann's anti-Hellenic argumentation, it may be that influences coming from life in the Greek world were positively creative in the formation of the very Pauline conception that he values so highly.

But before we go into the question of later interpretations, we should return to a more fundamental aspect. Even apart from Genesis, is it true for the Old Testament in general that death is something 'unnatural, abnormal, opposed to God'? Was there really, in biblical times, a 'Jewish connexion between death and *sin*' for Paul to 'presuppose'? In fact there are many powerful reasons against this understanding of the matter.

1. *God as author of life and of death.* Death in the Hebrew tradition, says Cullmann, is not something 'natural, willed by God'. Now there are indeed biblical materials that speak in this way. The God of Israel is spoken of as a 'living God', and the phrase 'as Yahweh lives' is a common formula in oaths.[4] 'With you is the source of life, and in your light we see light' (Psalm 36.10[9]). For humans likewise, 'life' is frequently spoken of as a supremely good thing: nothing is more to be prized than life. Jesus himself firmly aligned God with life: God is not the God of the dead, but of the living (Matthew 22.32; Luke 20. 38). Doubtless this powerful saying was in Cullmann's mind when he wrote the passage quoted above.

Even here, however, some caution is to be advised. I admit that I was personally surprised and impressed to learn that, apart from the oath formula 'as the Lord lives' (just over 40 times) or 'as I live' (23 times), the term 'living God' is quite rare in the Old Testament, with only fourteen cases or so, and limited in its scope.[5] Many or most of them appear as a formulaic expression in speeches of conflict against foreign opponents who have said something blasphemous against the God of Israel. According to Gerleman, God is mostly spoken of as giver or source of life, but to speak of him as 'living God' is done only with great reserve. In the Psalms, the place where one might be most inclined to expect it, the familiar and beloved cases of 'the living God' (at 42.3[2]: 84.3[2]) are more or less the only examples. Moreover, the noun 'life' itself is never used as a divine attribute, but generally stands for the result of divine action or blessing. Gerleman goes on to contrast this with other nations of the ancient Near East, which speak frequently and without qualification of the life and living qualities of their gods. If he is right, this is substantially different from what most readers of the Bible would have expected. And I do not see how Gerleman's learned analysis can be faulted. Some reserve at this point, therefore, is to be advised.

The Genesis story of creation, it is true, does seem to say nothing about death. God created living things, but there is no word of their dying. 'God did not create death', said the Wisdom of Solomon 1.13, 'nor does he rejoice in the destruction of living things; for he created all things for Being (εἰς τὸ εἶναι)' – all this, however, being part of an argument for *immortality*. Certainly Genesis 1 says nothing about God's creation of death. Yet perhaps silence about death does not mean that death was not part of the world as created? Darkness was not created, yet darkness was part of the created world. Possibly death had the same kind of status. Are we to suppose, on the basis of Genesis 1, that all creatures created by God were to live for ever, that there was to be no death at all in that universe? Certainly, if all the creatures were to live for ever, the world would quickly have come to be full!

No, for actually the Bible itself expresses another side. That God brings death is a familiar enough conception in the Bible. In Deuteronomy 32.39 the God of Israel, boasting his difference from any other gods, asserts that 'I kill and I make alive', a phrase that recurs several times and was obviously much appreciated. Hannah in her song (I Samuel 2.6) asserts that the Lord kills and brings to life. In Psalm 104.29f., within a poem of divine sustenance of the world that is doubtless prior in conception to Genesis 1, we hear how all creatures depend on God, and:

> When thou takest away their breath, they die
> and return to their dust
> when thou sendest forth thy Spirit, they are created.

We have here the same antinomy that occurs on another level with the question of good and evil. Obviously, just as God creates only life, and not death, he does only good, and not evil. But, in fact, other passages make it clear that he does evil too. 'Does evil befall a city,' asks Amos, 'unless the Lord has done it?' (Amos 3.6). Isaiah 45.7 speaks of God as the one who:

> forms light and creates darkness
> makes goodness but creates evil.[6]

There was a side of the world which Genesis 1 ignored, or perhaps deliberately eliminated from its picture, but which other sources expressed. Most obviously, if we may try to synthesize these opposing conceptions, let us put it in this way: God is supremely the giver of life, but life does not mean life without limit. Normally God gives life for a proper period, and to ancient Hebrews this would have seemed obvious. At the end of this proper period he takes it back. It is quite impossible to praise the 'realistic' view of life and of death in ancient Israel and at the same time suppose that the ancient Israelites thought death to be totally contrary to the will of God. The bringing of death was one of the major features of the character of this God.

In Psalm 104, then, God brings life to an end: that is part of the normal process. And when the Psalm adds that 'when thou sendest forth thy Spirit, they are created' (implying 'created again'?), this is not a reference to resurrection. The psalmist is talking of animal life: God does not resurrect the animal which has died; rather, he gives life to another in its place. Hannah's song may perhaps go somewhat further: it speaks much less of normal processes, much more of an overturning of the normal and its replacement by exceptional acts of the Lord, and if, in v. 6b, God not only 'brings down to Sheol' but also 'raises up' from there, we can agree that the text envisages even the giving of life to the dead. But Hannah's song is a psalm of praise rather than a cool statement of what normally happens. It goes far beyond what could count as directly relevant for a woman in her position, and it is not surprising that it was later taken to be a 'prophetic' utterance and filled with much extra content about the future.[7] It does not speak about what is normal, rather about the exceptional. But, in any case, it precedes and couples anything it says about God as giving life with the statement that he brings death; indeed in all three poems God's terminating of life comes before his fresh giving of life. That God limited life, and brought life to an end, was normal. It is quite wrong to suppose that death was entirely antithetical to the will of God.

2. *Death as completion and fulfilment.* As I have already mentioned briefly, there is an important side of Old Testament thought which appears to consider death in good circumstances – and this means, in particular, death in old age, surrounded by children and grandchildren, at the end of a good life, amidst honour and esteem – to be entirely proper and satisfying.[8] Thus the end of the book of Job tells how that patriarch, after all the vicissitudes related in the book, lived 140 years and saw his sons and grandsons over four generations, and died 'aged and full of years'; and it is impossible to doubt that this was a happy ending, the happiest ending there could be, for

a man who had been just and devout, had undergone bitter tests and come out well, and had been justly and fully rewarded for it all. This death was entirely in accordance with the will of God; indeed it is a powerful culminating sign of God's satisfaction with all that has happened. And Job is by no means exceptional in this. Abraham, certainly one of the great personalities, 'breathed his last and died' at the age of 175, in 'good old age' and 'was gathered to his people' (Genesis 25.8). There was, in ancient Israel, such a thing as a good death. It simply is not true, certainly not for the Old Testament taken universally, to say that death was always opposed to God, or was a curse, or that death was universally understood as a consequence of sin. A good death, in good circumstances, in old age and with descendants around one, was both right and proper, and formed a satisfying fulfilment and culmination to one's life.

And in Christianity itself, surely, we still maintain this insight: when a man or woman grows old and dies peacefully, their families around them, their life's work complete, their faith in God calm and untroubled, we do not say in our funeral liturgies that this ending of life manifests the final power of sin and its destructive effects. Of course we do not. It is a good death. And in ancient Israel it was so also. Such a death is not 'unnatural', it is not 'against the will of God', and surely few think so.

3. *Proper burial.* Death as the completion and fulfilment of life is dependent upon decent and proper burial.[9] This was the case in other lands and cultures also. The first person Odysseus spoke with in the underworld was his comrade Elpenor, who had been left unburied as they escaped from recent dangers: and his ghost at once beseeched Odysseus to provide a seemly burial, which was done. In Israel this is deeply related to kin and land. Nothing is a worse outrage, offensive to God and man, than that a corpse should be left unburied, 'with the burial of an ass', as Jeremiah said of Jehoiakim (Jeremiah 22.19; cf. 36.30, where we hear of the same king that he will have no descendant as successor, and his dead body 'shall be cast out to the heat by day and the frost by night'). To die in a foreign territory, equally, is a bad thing. Among his acts of charity Tobit gave primacy to the burying of any Hebrews whose bodies had been thrown out and left unburied (Tobit 1.17–19; 2.3–9). Burial in the proper place was again a sign of completion, of satisfaction of ancient wrongs.

The supreme example lies in the deep interest taken in the bones of Saul and Jonathan, disgracefully treated by the Philistines, retrieved as an act of devotion by the men of Jabesh Gilead for whom Saul had done a great act of deliverance, and finally brought by David to rest in the family tomb, an accomplishment of final atonement with which God was well satisfied (I Samuel 11.1–15; 31.8–13; II Samuel 21.12–14). Burial was more than a simple disposal of the corpse: it reunited the person with kin and ancestors. Such expressions as 'he was gathered to his kinsfolk' (used in Genesis 25.8,17; 35.29; 49.33; Deuteronomy 32.50 of Abraham, Ishmael, Isaac, Jacob and Aaron/Moses) surely meant more than that the corpse was placed in the same grave in which ancestors had been placed; similarly 'he slept with his fathers', an expression used of most of the kings, except where they were seriously disapproved of (and sometimes even when they were) or were killed by violence. It is reductionistic to consider these as mere constatations of the fact of physical burial in the family tomb. They are expressions of completion, fulfilment and satisfaction. A life had come to its end and its achievement, good, bad or indifferent, was fulfilled. It was not merely that the person had joined his ancestors in total extinction: in some sense he had gone to join them, and for God this was right and proper.

4. *Name and offspring*. As has been implied above, the presence of children and descendants is of first importance in the Israelite attitude to life and death. Through them the life of the individual is extended into the future.[10] The destruction of the 'name' is a painful outrage; the maintenance of the 'name' is of primary importance. Laws like those of inheritance and of the levirate marriage support this principle. The extension of inheritance rights to daughters of a family (Numbers 27; 36) is justified in order to prevent the extinction of a family name. Concepts of the promise of land are related to this. I follow Kellermann in quoting Ben Sira 41.11–13:

> Mourning for a man goes with his body
> but a name for graciousness will not be cut off;
> be fearful for your name, for it will accompany you
> more than thousands of treasures.
> The goodness of one living lasts only [a number of] days
> but the goodness of a name lasts for days unnumbered.[11]

The idea of the remembering of a name will prove to be an important point of origin for further development.

5. *Sheol*. Fuller discussion must be devoted to the concept of Sheol, the 'Hades' or underworld, the abode of death or of the dead. How

does Sheol fit into all this? It is not very clear how the modern discussion has dealt with it. The impression one gains is that Sheol is a blind alley. It does not constitute immortality, and it does not really lead towards resurrection either. Its value seems to lie in its negativity, which is thought to prove that existence without bodily life is no life at all. It is an attenuated, ghostly life, hardly worth calling life. Thus Cullmann speaks of 'a shady existence without the body, like the dead in Sheol according to the Old Testament, but that is not a *genuine life*'.[12] He uses the example of Sheol in order to show that there is no real life without a real body, and to point the contrast with the 'Greek' view that 'it is precisely apart from the body that the Greek soul attains to full development of its life'.[13]

Yes, Sheol is an uncertain area of obscurity in all this discussion. The Hebrews were vague about it. Yes, indeed, the dead went to Sheol: but did they all, universally, go there? Although Sheol is the abode of the dead, it seems to be spoken of mainly in connection with the persons disapproved, the evildoers, the ones who after a rich, powerful and successful life had to be cast down to the lowest and worst of states, or else in connection with quite wrong and disapproved things that happen to good people. The mighty kings were brought down to Sheol (e.g. Isaiah 14.4–15), and so were all the rich and powerful. But did Abraham go to Sheol, or Isaac, or Moses, or David? No one ever says so. Though there is no formal demarcation on moral grounds, actual talk of Sheol generally attaches to sinister characters and sinister events. David, giving his grim last testament to Solomon, makes it clear that Joab and Shimei should not be allowed to go down to Sheol in peace: these, regarded by David as bloodguilty villains, were the sort of people one properly associated with Sheol. But to take, for example, Saul: though he got into much trouble, no one talks of him going down to Sheol. He was, after all, the anointed of the Lord. Samuel was raised up after his death by the Witch of Endor, and the details sound very much like someone coming up from Sheol, but that expression is not in fact used of him. David, mourning for his lost son by Bathsheba, says, 'I shall go to him, but he will not come back to me', but he does not actually say that the child has gone to Sheol nor that he himself will join him there. If Jacob says, after Joseph has disappeared, 'I will go down in mourning to Sheol to my son' (Genesis 37.35), this may be precisely because his son (as he supposes) has died a violent and untimely death and he himself will die an improper death, having been thus deprived of his son. These things are said in speech by the characters of the narratives, but the narrator himself holds back from committing himself. No doubt it was thought that Sheol was the universal destination of the dead, but the deaths of great Hebrew persons were not so narrated. There was a reserve about speaking so of the most hallowed persons. And on the other side this negative perception may be one starting point on the route which was to lead eventually towards the idea of hell, a place uniquely for the evil. So, to sum up this point, though Sheol was the place for the dead, there were nuances in its character and implications which may have led in creative directions.

Sheol, it seems, is not something derived from the initiatives of biblical, Yahwistic religion: it was something that was inherited from an ancient past, something that was already there, a part of 'natural religion' in that sense. Nothing we know from Yahwistic prophets or other traditionists could be said to have created or originated the concept of Sheol. It is not surprising, therefore, if the Homeric picture of the underworld in the Νεκυία of Odyssey 11 serves in many respects to fill out the very limited Hebrew depictions of this sad territory. This is important, for it means that this earlier stage of Greek material, and one highly comparable with the Old Testament, actually shares important insights with the latter.[14]

Certainly, the survival of the shades in Sheol was not seen as in any way an overcoming or transcendence of death. Sheol was not a 'life after death' but was part of the reality of death itself. Existence there was very miserable and scarcely deserved to be counted as 'life'. These are the valid points in the arguments we have been describing. It may nevertheless be the case that Sheol has a significant part to play in the development of ideas of immortality.

First, Sheol, though certainly a miserable existence and not a life in any full sense, was still a sort of *continuance*. Persons were not merely or totally extinguished at death. This fact was not, as I have said, an antidote to the fact of death or an amelioration of its seriousness; on the contrary, it was a part of its seriousness. Nevertheless, it was still a sort of continuance. When David said of his dead child 'I will go to him' it meant that there was, in however shadowy a form, a 'him' for him to go to join. Moreover, it seems to have provided – at least if we may take the Homeric parallels as a guide – one of the essential elements which is continually mentioned in philosophical discussions of immortality: it provided continuity of *identity*. Achilles in Hades was still, in some sense, the same person who had earlier excelled in the great battles around Troy. He could speak.[15] He could be asked how things were, down there in Hades, and could reply, as he did, with a very negative answer. There was a continuity of character as there was of appearance. Persons could be recognized. Odysseus knew his mother. Though there was no bodily substance, and when he embraced her there was nothing there for him to hold, her face and shape were the same as they had been during life. The shades were recognizable. Samuel, apparently, could be recognized by his cloak and general mien. His shade uttered the same kind of angry tirade against Saul that had characterized the same seer during his earthly life. The shades could talk about their past life and compare it with the miserable existence now inflicted upon them. They worried about what had happened to their children and complained about injustices they had suffered; they insisted, particularly, on the importance of their own proper burial, if that had not yet been carried out. Sheol, then, was not total extinction.

Now in the Old Testament this was very important, at least in poetic expressions. The question was not, primarily, whether Hebrew persons continued to exist, either in a full sense or in a very etiolated sense, but whether they continued to be in touch with Yahweh. One position on this, best expressed in the ominously beautiful Psalm 88, was that Yahweh forgot them:

> My life draws near to Sheol,
> I am reckoned among those who go down to the Pit;
> I am a man who has no strength,
> like one forsaken among the dead,
> like the slain that lie in the grave,
> like those whom thou rememberest no more,
> for they are cut off from thy hand (Psalm 88. 4–6 [3–5]).

But the Psalmists, or some of them, were not all content to leave it at that. Even the poet of Psalm 88 was still praying that God would no longer 'cast him off'. And others went further and uttered the complementary belief that Yahweh, being the sole God and the God of all areas of existence, had power in Sheol also, at least potentially, or could be present there:

> If I ascend to heaven, thou art there!
> If I make my bed in Sheol, thou art there! (Psalm 139.8)

This belief stems, doubtless, from the basic Hebrew belief in the One God. In Greece there were the gods of the underworld, who might not take much care for the inhabitants of that region, but at least their existence meant that the gods of normal upper existence were absent from this region. In the Old Testament, at least in the approved religious currents, such deities were eliminated. The cult of the dead, necromancy, the sacral aura surrounding death, were restricted or eliminated. The world of death was made *tabu, tabuisiert* (Kellermann). There was either Yahweh or no god at all. But Yahweh claimed dominion over all areas. His power must therefore extend even into the gloomy reaches of Sheol. Sayings to this effect can be quite old; I have already quoted Hannah's Song, and consider again:

> Though they dig into Sheol [i.e. to escape God's judgment]
> from there shall my hand take them;
> though they climb up to heaven,
> from there I will bring them down (Amos 9.2).

Such a passage is in one way metaphorical: it is saying that there is nowhere to go in order to escape from God. It does not deal directly with life after death. But it is still an indication of a significant possibility: God can exercise power in Sheol.

The real questions which pious Israelites occasionally raised were two:

1. Is there contact with Yahweh in Sheol, are there means to praise him and seek him?

> Whither shall I go from thy Spirit?
> or whither shall I flee from thy presence?
> If I ascend to heaven, thou art there!
> if I make my bed in Sheol, thou art there! (Psalm 139.7f.)

2. Since Sheol is so ghastly a place, is it conceivable that God will leave his devoted ones there, that he will not be able to keep them from the dismal darkness and negativity of the Pit?

> Therefore my heart is glad, and my honour rejoices,
> my flesh also dwells secure;
> for you do not give up נפשי, my *nephesh*, to Sheol,
> and you will not let your devoted one see the Pit (Psalm 16.9–10).

Now these utterances are not doctrinal statements that might define the nature of the 'afterlife'; they are doxological terms of praise and worship, and do not explain *how* the 'soul' is to be kept away from Sheol or what God actually does when he comes to be present in Sheol. Nevertheless, they constitute a declaration of importance: the God of Israel has, potentially, presence in Sheol, power to control the destiny of his own ones who are there, but, most important, ability to hear their prayers and to have some sort of communion with them. Moreover, and as a result, he has power to keep them out of Sheol and, if need be, to remove them thence.

These thoughts help us to understand other poems, such as Psalm 49. This psalm is very aware of the problem of death. In particular, the wise die just as the foolish do (v. 11[10]), and humanity is like the beasts which perish (נמשל כבהמות נדמו 13[12]). Death is the portion of those who are foolishly self-confident (there is much unclarity in the exact wording here), but it is clear:

> God will ransom my soul;
> from the hand of Sheol indeed he will *take* me (my own rendering)
> ... (v. 16[15]).

It is very likely that the word 'take' here is modelled on the idea of God's 'taking' of such as Enoch in Genesis 5.[16] The same occurs in Psalm 73.24:

> You guide me with your counsel
> and afterwards you will *take* me in (to?) glory.

In phrases of this kind the known terminology of God's 'taking' of exceptional persons, inherited from ancient stories, comes to have a part in Israelites' outpouring before Yahweh of their personal sense of need and limitation.

Another aspect, much emphasized especially by the scholarship of Christoph Barth, is the idea of death as a powerful domain which can intrude into this life. The realm of death expands and seizes people through the power of sin, through sickness, through hostility and injustice, perhaps through magic and other chaotic forces. 'Death' in this sense is not to be defined by what we call 'physical death'. It is a cosmic power which can seize hold of us while we are still walking about, talking, acting. The reality of these concepts has been much discussed,[17] and they are extremely important especially for the understanding of Hebrew poetry. However, for the discussion of such questions as whether for biblical man death was 'natural' or not, we have to observe the existence of a semantic gap between what *we* are meaning and what the biblical language is referring to. For when a biblical poet in sickness or in deep trouble thinks that he is being brought into the realm of 'death' or 'brought down to Sheol', or that weakness of life is a form of death, we have to recognize that 'death' in such terms is not the same thing as what we call death.[18] It is not so much that 'death' is used in a metaphorical sense, as when in modern times one says 'I'm absolutely dead' (e.g. from exhaustion on a shopping trip); it is rather, if we follow Barth's understanding, that the conceptual boundaries of 'death' are serious but differently defined. There is a difference between this kind of biblical 'death' (with inverted commas) and what we mean by death when we discuss whether death is natural or not. If we were to apply the biblical usages univocally to our own life, then all the patients in the emergency ward of a hospital would be 'dead', and so no doubt would be all those undergoing psychiatry for depression; only when moved into the convalescent ward, or when released from psychiatry, would they again have 'life' and be no longer 'dead'.

If it is true that this correctly represents the language of the Bible (and I am still not sure that it does), all it seems to prove is that biblical materials framed in this language are not suitable for helping us with problems of what *we* call death. 'Death' in this poetic biblical sense may be a curse, may be opposition to God, may be a force that challenges him and opposes him, but that only proves that we are talking about something other than death. If, let us say, a hypothetical biblical person, using this idiom, should say 'I was dead but God made me alive', he would be meaning something like 'I was suffering terrible pains, terrors and isolation, but God restored me to health, freedom from fear, and community'. This would be serious talk, but would still not be talk about what *we* mean by the term death.

We might compare the usage of 'revive' in our own speech: though this is derived from 'make alive (again)', it is seldom if ever so used. It means rather 'to give vitality, liveliness' after fainting, unconsciousness, depression and the like. On this side the biblical חיה ('live') can be rather similar, especially in the *piel* and *hiphil* forms. Thus the *piel* can mean 'allow to live, i.e. to continue living' (e.g. Exodus 1.17, of the Hebrew baby girls); the *hiphil* can mean 'to restore to health (a leper)', II Kings 5.7, cf. of Hezekiah Isaiah 38.16; of vegetation the verb can mean 'cause to grow, produce (grain)', Hosea 14.8; it can mean 'revive (the heart and spirit, of the lowly and oppressed)', Isaiah 57.15 twice; and it can mean 'restore to life' (the dead), II Kings 8.1, etc. Thus in Hebrew the terminology of 'living' covers a considerable variety of somewhat different things, and this is so with 'reviving' in our usage. In Hebrew, if this scholarly trend is right, the same is true of the term 'death' – at least in the Psalms.

And yet, on the other side, one should not exaggerate the extent of this kind of vocabulary. In the vast majority of cases, when the Hebrew Bible talks of death, it means death as we do. At first sight, as was suggested above, this might simply mean that many cases of 'death' in Hebrew poetry are irrelevant to our questions of death, immortality and resurrection. However, on further examination this may not be so. For poetic expressions, as in prayers for deliverance from 'death', may indeed have been first meant to refer to a power that could seize and overpower the living, but the same formulations, later viewed from the customary daily sense of the words, could well have furnished a starting point that led towards expectations of deliverance from death in the normal and physical sense of the word.

The same may be true of some passages in which the opposition between God and 'death' may perhaps go back to an older situation where Mot (= 'Death') was a deity, locked in combat with another, a theme which has become increasingly important since the Ugaritic texts became known. Here again we have an opposition which is powerful and dramatic but which, if it is so interpreted, does not demonstrate a direct opposition between the God of Israel and death in our normal sense of the word. Once again, Yahweh may be battling with the god 'Death', but this only means that there is no clear and direct reference to what we call 'death'. In cases of this kind, we are dealing with remnants of ancient theomachic conflicts, and passages of this kind cannot be taken without further ado as evidence of Hebrew attitudes to life and death.

Nevertheless, as time goes on, poetic and prophetic expression seems to express more strongly the longing for the overcoming of death: so Isaiah 25.8; Hosea 13.14, and other comparable passages. Here again, what was originally a remnant of theomachic poetry may have come to express a belief that Yahweh was really hostile to death in its normal sense. But, as will be argued again shortly, some of this development may be part of the idea of a regained paradise as a realm of *peace*, and in that sense is a reaction not simply to *death* of any kind but to violence, to the consequent suffering, and in particular to *war*.

6. *Hebrew 'totality thinking' and the soul.* Equally important, in the modern study of these matters, has been the picture of the 'totality thinking' of the ancient Hebrews.[19] 'In Israelite thought man is conceived not so much in dual fashion as "body" and "soul", but synthetically as a unit of vital power or (in current terminology) a psycho-physical organism', wrote A.R. Johnson.[20] Undoubtedly these thoughts have exercised a strong influence on ideas about the nature of life and the character of death in biblical Israel. Westermann is quoted by Knibb as saying that 'a human being does not consist of a number of parts (like body and soul and so on) but rather is "something" that comes into being as a human person by a quickening into life ... a person is created as a *nephesh ḥayyah*; a "living soul" is not put into his body'.[21] Or, to use another popular formulation, man does not 'have' a soul, he *is* a living soul. Unquestionably these ideas have been very influential in our time.

Central to them is the term *nephesh*, often translated as 'soul' in the older biblical versions. Writers commonly give a prominent place to the creation of Adam (Genesis 2.7). When the first man was created, God fashioned him out of dust or mud and 'breathed into his nostrils the breath of life; and man became a living *nephesh*'. From this, and some other similar examples, the conclusion is drawn that *nephesh* means a total, indivisible, unity of the person, body, soul, and whatever else. The word, we are reminded, may mean a part of the body, like the throat; in Numbers 6.6 it may even refer to a corpse (על־נפש מת לא יבא: the Nazirite should not come upon the *nephesh* of one dead); it may thirst and be desperate. It is said to be emblematic of the thought of the Hebrew in totalities, widely disparate from the analytic study of the Greeks and of modern times. I have to confess to having said some of this myself at times: but, when one looks afresh at the material, is it true?

Can it all be true? There are so many reasons against it. Is it even remotely plausible that ancient Hebrews, at the very earliest stage of their tradition, already had a picture of humanity which agreed so well with the modern esteem for a psychosomatic unity? How did they manage to get it all so perfectly right, when the Greeks, apparently, so thoroughly misunderstood everything? Is there not an obvious bias in so many modern textbooks, which seem to want nothing more desperately than to deny that the Hebrews had any idea of an independent 'soul', worse still an immortal one? May it not be a mistaken semantic analysis, inspired by admiration for the very 'totality thinking' that it is supposed to demonstrate?

Suppose we take it in another way. The word *nephesh* has a wide variety of meanings. These overlap, but remain distinguishable, and there is no unitary '*nephesh* concept'. The choice between meanings is known through the verbal and syntactic contexts at each point. Only in some cases and in some contexts is *nephesh* relevant for the discussion of the 'soul'. And the 'soul', we may suspect, may occasionally be indicated by other terms also.

Let us return therefore to the supposedly key instance Genesis 2.7. *Nephesh ḥayyah* means 'a living being' or something like that: it is a common term for animals. Made out of mud, and breathed into by God with the 'breath of life', the first man becomes a living and moving, an animate, being. We can call that a totality if we wish. But for questions about the nature of the soul this is beside the point; for *nephesh* here does not mean 'soul'. In Genesis 2.7, the element which is closer to the soul is the 'breath' that comes from God, which is quite distinct from the mass of mud out of which Adam was made. Adam is fashioned of mud, and God breathes into him the breath of life; he thus becomes an animate being, like any other animal. The passage is obviously dualistic: there are two ingredients in man, the mud and the breath. Or, alternatively, one can make it monistic: then the 'breath' is not 'part' of man but something that comes into him and goes out again; the only element that really *is* man is the mud or dust. 'Dust you are, and to dust you will return' (Genesis 3.19) is the obvious continuation of this line of thinking. If there is anything in the verse that comes close to the 'soul', it is not the *nephesh ḥayyah* of the end of the verse but the 'breath' which God infused into man. What cannot be done is to argue, as countless presentations have argued, that the *nephesh ḥayyah* of the text tells us that the 'soul' of Hebrew thinking is the totality of the human person. The text says nothing to that effect. The argument results from the crude, unthinking, transference of the meaning 'soul', traditionally attached to *nephesh*, to this phrase where the word has a different meaning altogether. It is a simple fallacy:

(1) *nephesh* is the Hebrew word for 'soul';
(2) *nephesh* is the totality of Adam as a living being;

(3) therefore the Hebrew 'soul' is the totality of a living being, including something, or all, of its physical character.

The argument depends upon that which it denies: for it depends on the traditional idea that *nephesh* is the word for 'soul', which is what it is attempting to disprove.

It is the polysemy of *nephesh* that makes such argument impossible. The fault lies in the mistaken supposition that there is a 'concept' *nephesh* which is formed by the bringing together of all the various meaning-relationships that this word can undergo.[22] When we are talking about 'soul', we have to concentrate on those cases of *nephesh* which *do* mean 'soul', and perceive the possible irrelevance of the many cases where this word means something else. Westermann in his important article[23] rightly perceives that there is only a limited group of cases of *nephesh* where 'soul' represents the meaning; but he again rightly perceives that there are cases where this meaning cannot be avoided.

There are, then, other contexts in which it seems that *nephesh* is much closer to what has traditionally been understood as 'soul'.[24] For discussions of the soul we have to look primarily at these examples. The problem is that many cases are somewhat ambiguous: *nephesh*, let us say, might be assigned to the meaning 'life', but might equally be assigned to the sense of a (let us say, partially) independent 'soul'. It is obvious that the enormous modern cultural pressure in favour of an understanding as the 'totality' of the person has moved scholarship in that direction. Moreover, modern translations have tended to move away from the rendering 'soul' and thus obscure for the general reader the extent of the evidence. Often another interpretation would be plausible, if one had the energy and imagination to pursue it. And the fact that Genesis 2.7, often taken as the 'parade' example for the 'totality' reading, is certainly a false example, encourages one to look in other directions. Is it really true that there is never a 'soul' which is anything different from the psychosomatic totality of the person? I do not think I can prove, at present, that there is such a 'soul', but I think we can show that it is a reasonable possibility.

First we may consider those cases where the individual addresses his own *nephesh* (Psalms 42; 103): are we to suppose that the (total?) person is addressing his own totality?[25] I quite admit that there is a totality of the human person involved. But it is equally possible that the person is addressing the 'soul', which is something like a superior companion or accompaniment to that totality. Taken on the customary 'totality' interpretation, these passages seem rather silly. It makes more sense that he is addressing his soul, an entity which is associated with his psychosomatic unity but in some sense is also outside it, independent of it. I do not see why they could not have thought in this way.

Secondly, we have to look at those cases in which *nephesh* stands in some kind of opposition. A phrase like Isaiah 10.18, מנפש ועד בשר ('all the distance from *nephesh* to flesh'), is metaphorical in its context and is doubtless a proverbial saying; no doubt it has the general effect of 'completely'.[26] But it clearly depends on, and thus demonstrates, an opposition between *nephesh* and flesh: they may not, generally, be two totally separable parts, but they are – in instances of this kind – at the opposite end of the spectrum. And it is these instances that are of first importance in discussing the idea of the soul.

Cases in poetic parallelism may well be deemed ambiguous:

Job 14.22: His flesh is painful upon him and his *nephesh* mourns;
Psalm 63.2[1]: My *nephesh* thirsts for thee; my flesh faints for thee.
Psalm 84.3[2]: My *nephesh* ... my heart and flesh.[27]

Parallelism in itself does not prove whether the two elements are seen as closely associated, or as at the opposite ends of the spectrum. Even if we decide for close association, it can hardly be surprising if some prefer the opposite. One cannot be sure, but I think that the parallelism is probably an oppositional one: the flesh and the soul are very different entities. Precisely because they are here doing the same thing, the poetics are effective.[28]

Important evidence also lies in the apparent *mobility* of the soul under certain circumstances. There are cases where the *nephesh* 'leaves' a person, and, sometimes, 'returns'. When Rachel died (Genesis 35.18), her *nephesh* went out from her. Here again the text may be ambiguous. Does it mean 'as her life went out', in which case nothing about the soul as such is said? Or does it mean 'as her soul left her', in which case one may ask: where then did her soul go, and what happened to it? RSV renders 'as her soul was departing'; REB sternly demythologizes to 'with her last breath'. If the *nephesh* is just a life-force, or simply means 'life', then the passage tells us nothing about the soul involved; if it means the soul, then it tells us a lot. For it suggests that the *nephesh* is thought of as something that lives along with, or in, the person, but has a certain independence and is the primary life-giving source of potency within the person. When it leaves, of course the person dies. In view of the enormous evidence of such ideas of the 'soul' in ancient peoples, I cannot see why this reading of the text must be ruled out. May I take the liberty, from now on, of calling it 'soul'?

It is the same with the child of I Kings 17.21f. The child's 'breath' had ceased (v.17). Elijah prayed that his *nephesh* should return to him again; and it did return into him (literally 'into his insides') again, and he revived. Why can this not be understood as a soul which departed, and then returned? Again, perhaps, it cannot be proved, but to me it seems quite likely.

The soul seems to extend itself beyond the frontiers of the person. Jonathan loved David 'as he loved his own soul': he did not just love himself, he had a soul which it was meaningful to love. When Jonathan's soul was 'tied' or 'knit' or linked with the soul of David (I Samuel 18.2), this is more than a metaphorical expression for their affection: the soul reached out to meet with that of David and joined itself with his. In Song of Songs 5.6 the singer's 'soul went out' to her lover when he came to the door and spoke (REB 'my heart sank'). This extension and mobility of the soul is an important reality which should not be explained away.

The association of the 'soul' with life has been rightly emphasized by all researchers in this area. But does the *nephesh* itself die? Since the *nephesh* is so closely associated with *life*, it seems unlikely that the *nephesh* is normally supposed to die, and, in spite of impressions sometimes given to the contrary, the texts very seldom say that it does. There are instances, but the cases are so exceptional as to prove the point: Balaam in Numbers 23.10, and Judges 16.30, where Samson says 'let my *nephesh* die with the Philistines', similarly also Job 36.14. However, in narrative we seem never to have a text that says 'his (or her) soul died'. Over the hundreds of narratives of deaths this is striking evidence. If there had been, let us say, twenty or thirty cases which said 'his *nephesh* died', we would have been told that this was definitive evidence that the *nephesh* was the totality of a human person, and excluded all ideas of an 'afterlife'. But there are few or no such cases. It would be conceivable that the soul might –

doubtless with divine support – extend itself in time beyond bodily death. In Israel, events of this kind did not have to be synchronous. In I Samuel 25.36–38 Nabal hears terrible news from his wife Abigail, and 'his heart died in his insides and he became a stone'. But this did not finish him off, for it was about ten days later that the Lord smote him 'and he died'.

In saying that the soul might extend itself beyond bodily death, I am not opting for 'immortality'; there are other possibilities: for instance, that the soul continued to live as an independent entity for some time, or under special conditions. (A rabbinic thought along these lines will be mentioned shortly.) If so, it did not necessarily die when the body died by a natural death, though it could be put to death or destruction by hostile forces, such as Sheol. Thus there are a few cases of 'you have delivered my soul from death' (e.g. Psalms 33.19; 56.14[13]; 116.8; cf. 78.50). These might be classified in this way.

To this may be added the close associations of *nephesh* with *ruaḥ*, commonly translated as 'spirit'.

> In my soul I yearn for thee in the night; in my spirit within me I seek for you at dawn (Isaiah 26.9).

> The bitterness of my soul/ the life of my spirit (this last a passage concerning Sheol, Isaiah 38.15–16).

Though *nephesh* may be – sometimes – the psychosomatic totality of a human being, this does not readily work with *ruaḥ*. *Ruaḥ* is likely to be in systematic opposition with *baśar* ('flesh'): in this portion of its usage, it is something that is movable and free from fleshly involvement, though it can of course involve itself at times. Cases of this kind further support the understanding of *nephesh* as markedly distinct from 'flesh'.

From the point of view of those discussing resurrection and immortality, however, perhaps the most powerful observation is this: that in the passages concerning the possible leaving of the person in Sheol, there is a marked use of נפשי, 'my soul', and a marked predominance of this over the other anthropological expressions that might accompany it or be in parallelism with it.

> You have brought up my soul (so RSV) from Sheol (Psalm 30.4[3]).

You have delivered my soul (RSV) from the depths of Sheol (Psalm 86.13[12]).
You have held back my soul (RSV life) from the pit of destruction (Isaiah 38.17[18]).
My soul is full of troubles and my life draws near to Sheol (Psalm 88.4[3]).
You do not give up my soul to Sheol (Psalm 16.9[10]).

The poets seem not to say: you do not leave my heart, or flesh, or liver, or kidneys, in Sheol. It is about the soul that the question is put.

Here again the Homeric Hades may be compared. When Odysseus tried to embrace his mother and found that there was nothing for him to grasp, he feared that Persephone had sent forth a mere εἴδωλον, but his mother explained that this was what normally happened at death: the sinews no longer have flesh and bones, which are burnt away after the θυμός has left the bones, and the ψυχή flies away like a dream (Odyssey, 11.210–224). In this aspect, the Homeric view was also a 'totality' viewpoint. Flesh, bones and the like had decomposed, and the ψυχή *was* the totality of what continued to exist.

As we have seen, the cases which may mean 'soul' are often somewhat ambiguous. 'Soul' is not distinguished by an easy terminological marking from other meanings of *nephesh*. The cases that mean 'living being', 'animal', 'person' (as with numbers, in lists, etc.) are readily identifiable from their syntactical situations. This makes it seem easier to emphasize these latter cases, while the cases which may potentially mean 'soul' can easily be made to disappear through the pressure of ambiguity. Another misleading aspect has been the false illusion of objectivity caused by etymology, and by those linguistic methods which take physical meanings as more 'basic' than others. These considerations have drawn attention to the meaning 'throat' for *nephesh*, a meaning which seems to me to be quite marginal for the general understanding of the Bible. Thus a study like Wolff's *Anthropology of the Old Testament*, out of sixteen pages on *nephesh*, devotes only one to 'soul', and precedes it with four on 'throat' and 'neck', while adding three on its use where 'my soul' is said to be no more than a way of expressing the personal pronoun 'I'.[29]

I submit, then, that it seems probable that in certain contexts the *nephesh* is *not*, as much present opinion favours, a unity of body and soul, a totality of personality comprising all these elements: it is rather, in these contexts, a superior controlling centre which accompanies, expresses and directs the existence of that totality, and one which, especially, provides the life to the whole. Because it is the life-giving element, it is difficult to conceive that it itself will die. It may simply return to God, life to the source of life. Otherwise, it may still exist, and the thought of its being brought down to Sheol, or being killed, is intolerable. It is *particularly* the thought of the envelopment of the *nephesh*, in the sense of 'soul', in Sheol that is hateful. With the recognition of this fact the gate to immortality lies open. I do not say that the Hebrews, in early times, 'believed in the immortality of the soul'. But they did have terms, distinctions and beliefs upon which such a position could be built and was in fact eventually built.

Incidentally, it is interesting to reflect that Johannes Pedersen, whose great book *Israel* was in many ways the most profound study of Hebrew totality thinking, was in no way inclined to reduce the soul to something like human psychosomatic life: on the contrary, he magnified the importance of soul, filling the whole world with souls, and making body into a kind of soul rather than the reverse.

In spite, therefore, of the importance of Hebrew totality thinking – which I do not dispute in principle – some formulations of the Hebrew Bible point towards more dualistic conceptions. As we have seen, God's formation of Adam, 'dust from the earth', and his breathing 'into his nostrils the breath of life' (Genesis 2.7), in spite of all that has been written, look awfully like two ingredients, of which Adam was a sort of compound. Take away the breath of life, which comes from God, and the body returns to dust.

The later writer Qoheleth, doubtless thinking over this same theme, thought of death as just such a dissolution: 'the dust returns to the earth as it was, and the spirit returns to God who gave it' (Qoheleth 12.7). That the spirit returns to God who gave it could perhaps mean simply that the life-giving breath was withdrawn and man 'expired': in other words, the 'breath' was a sort of impersonal force which held the bodily elements together, constituted life, and was eventually withdrawn. But it could mean, or could eventually be taken to mean, that that breath was more like the 'spirit' of our terminology and that the spirit of the person returned to God from whom his or her life had come – a thought that parallels the mode in which in sacrifices the blood, which is life given by God, returns to

God who is its source. Such conclusions may not have been those of the Genesis text in its original context, but one can hardly be surprised by the fact that they were probably drawn.

In fact, we can go farther. If the saying of Qoheleth had in mind the story of human creation in Genesis 2.7, or at least the same ideas, which I think likely, his use of the term *ruaḥ* ('spirit') seems a highly significant change: for this word could not bear the highly physical components which people have attached to the 'totality' concept of *nephesh*. He meant the *human spirit* which went in a different direction from the flesh which returned to dust. The ancient Hebrew conceptions of humanity were such that conceptions differentiating body and soul could be built upon them, already within later Old Testament strata – as, in later Judaism and early Christianity, they certainly were.[30]

This is the other side: even if we grant the maximum to the totality thinking of the earlier Hebrews, in later times dualisms of various kinds became very strong. The Dead Sea Scrolls are a central area for the study of these. Otzen in a fine survey distinguished 'a *psychological-ethical dualism* (the nature of man is constituted of two opposite powers or "spirits"), a *cosmic-ethical dualism* (man and the world divided into two groups led respectively by the "prince of light" and the "angel of darkness"), and an *eschatological dualism* (the present world, under the rule of Beliar or the "angel of darkness", will be succeeded by a new world under the dominion of God)'.[31] The 'two ways', light and darkness, the two inclinations in the heart of man, add to the collection. To these, undoubtedly, are to be added body (or flesh) and soul, mortality and immortality. Josephus, describing the Pharisees, says quite clearly of them that they maintain that 'every soul is ἄφθαρτον, immortal or imperishable, but the soul of the good passes into another body, while the souls of the wicked suffer eternal punishment'.[32] Of the Essenes he says that they believe that 'the bodies are corruptible ... but souls are to be immortal and continue for ever' (φθαρτὰ μὲν εἶναι τὰ σώματα ... τὰς δὲ ψυχὰς ἀθανάτους ἀεὶ διαμένειν). It is impossible to brush aside the evidence of these distinctions within later Jewish and early Christian writings. What was Jesus talking about when he told people not to fear the one who would kill the body, but to fear the one who could kill both the body and the soul in hell (Matthew 10.28)?[33] Humans could kill the body but not the soul; only supernatural powers could damage or kill the latter. His words are in entire continuity with the writings of the Maccabean crisis of martyrdom, of which I shall speak in a moment.

Writers convinced of the 'totality' picture of Hebrew *nephesh* have often, logically, followed this up with the assertion that the rendering of the word by Greek ψυχή distorted the meaning. Westermann is right to note the counter-argument of Bratsiotis to the effect that this judgment cannot stand if the pre-Platonic Greek usage is taken into account.[34]

Hebrew thinking, then, is not itself a monolithic block. Even if it has been correctly described for the earlier biblical period, any picture based on that period requires to be greatly modified for later times. And the Greek side of the comparison is equally in need of qualification. Theologians who have used the contrast between biblical thought and Greek thought have commonly accepted a picture of Greek thought built upon the Platonic philosophical tradition; but it is easy to see that the Greek picture of life and death in earlier times may have much more in common with that of early biblical times.[35] Hostility to Greek thought has probably caused important evidence, in which the early Greek development can be suggestive for the Hebrew realities, to be neglected. Likewise, but conversely, even if the thought of ancient Hebrew texts is widely divergent from that of Greek philosophy, it is likely that some of the later and post-biblical Jewish texts have much more in common with the latter. The accepted contrast between a Platonic philosophical picture and a Hebrew totality concept may be constructed upon very limited sectors and periods on both sides, arbitrarily selected and inadequately investigated.

A moment may be taken to remind the reader that in later (Talmudic and post-Talmudic) Judaism there was a rather vague and fluid combination of resurrection of the body and immortality of the soul. For some examples I quote from the article 'Afterlife' in *Encyclopedia Judaica*:[36]

> When a man dies his soul leaves his body, but for the first twelve months it retains a temporary relationship to it, coming and going until the body has disintegrated. Thus the prophet Samuel was able to be raised from the dead within the first year of his demise. This year remains a purgatorial period for the soul, or according to another view only for the wicked soul, after which the righteous go to paradise and the wicked to hell. The actual condition of the

> soul after death is unclear. Some descriptions imply that it is quiescent, the souls of the righteous are hidden 'under the Throne of Glory' (Shab. 152b),[37] while others seem to ascribe to the dead full consciousness (Ex.R. 52.3; etc.). The Midrash even says: 'The only difference between the living and the dead is the power of speech.' There is also a whole series of disputes about how much the dead know of the world they leave behind (Ber. 18b).
>
> In Maimonides ... the immortality of the soul is paramount. Though he makes the belief in resurrection, rather than in the immortality of the disembodied soul, one of his fundamental principles of Jewish faith ... it is only the latter that has meaning in terms of his philosophical system. Indeed, the resurrection does not figure in the *Guide of the Perplexed* at all.

Take again the case of Spinoza. One of the suggested reasons for his excommunication, in Amsterdam on 27 July 1656, was that he thought that 'the soul dies along with the body', a shocking opinion amounting to a form of sheer atheism. Commenting on this, a recent article remarks that: 'The main theological role of the belief in the immortality of the soul is to establish the foundation of reward and punishment not in this world.'[38] If this was a factor in the excommunication of the distinguished Spinoza, it is clear that immortality of the soul was a serious matter in Jewish life.

Well, the reader may say: this may be so, but of course it is only because these Jewish traditions had imported the concepts of Greek philosophy. Yes, perhaps. But this is just the point. People have not only been using a crude and questionable opposition between Hebrew and Greek thought, but they have been implying that the Hebrew thought, thus identified, is *perfect and complete*. As it is depicted by many writers, it leaves no problems, no insoluble dilemmas; it contains no contradictions and it answers all the questions. This being so, if anyone was attracted to elements of Greek thought, it is because they were fools or knaves. Having a perfectly adequate mode of thought, they were willing to spoil it through the introduction of faulty and inadequate ideas from Plato or others, ideas which could only wreck the entirely satisfying totality that already existed.

All this has been illusion. Even if Hebrew 'totality thinking' has been rightly described, which may well be doubtful, it left gaps, antinomies and contradictions, the evidence of which can be seen in the later stages of interpretation. These later stages were busily changing their ground in order to meet some of the consequent difficulties. If post-biblical Jewish sources are interested in Greek ideas and find them creative, it is because they had problems in their own religious tradition which they found to be at least partly alleviated through conversation with these ideas and, where desirable, partial adoption of them. Very often, if Jews adopted Greek concepts, it must be because they thought that these expressed very well the concepts of their own ancestral religion. Lifschitz quotes a poignant Jewish inscription from Caesarea: ἔσηται *(sic)* ἡ ψυχὴ ὑμῶν ἐχομένη ἀθανάτου βίου ('your soul will be in grasp of immortal life'), and points out that it is a rewording of I Samuel 25.29: καὶ ἔσται ψυχὴ κυρίου μου ἐνδεδεμένη ἐν δεσμῷ τῆς ζωῆς ('and the soul of my lord will be bound up in the bundle of life').[39] The biblical original speaks out of the older Hebrew world, but the Caesarean version was, to them, a very fine restatement.

The same, we may suspect, was true in Christianity: even if the New Testament is wholly concentrated on the resurrection of the body, that concept itself had many problematic areas, and that same concentration on the one concept may have been a powerful reason why so many problematic areas remained to be examined.

7. *Later Hebrew thinking*. As we have seen, the way in which St Paul uses the story of the garden of Eden depends in large measure not upon the Genesis story in itself but upon the way in which that story came to be handled and interpreted in later times, and especially in the Hellenistic era. I cannot attempt in a limited space to examine all the relevant developments, but some aspects have already been mentioned, and others must now be added.

In the later stages of the Hebrew Bible some changes of emphasis seem to occur. I remarked already that in the entire Old Testament there is no reference back to the story of Adam and Eve in order to explain or account for sin or evil. But by later times this is changing, and in what is now the deutero-canonical literature there is an efflorescence of interest in the narratives of primeval times.

One symptom of the change may be seen in Ezekiel's picture of the king of Tyre in ch. 28. It is impossible here to disentangle the many textual and interpretative problems, but, very much simplified, we may say that there are two units, and in the first, Ezekiel 28.2–10, the king is definitely seen as a human who imagines for himself the status of the divine. He has wisdom, and has had great success in commerce and gathered great wealth. However, because he imagines himself as a god, he will be assaulted by aliens, who will thrust him down to the Pit of Sheol.

The second oracle, 28.11–23, depicts the king of Tyre as 'the signet of perfection', who was in Eden, the garden of God, among the precious stones, on the holy mountain of God. He was blameless in his ways from the time of his creation, until iniquity was found in him, and then he was cast out from the mountain, spoiled and reduced to ashes by fire. There is mention of a 'cherub', which reminds us of Genesis 3, but textual problems have made scholars uncertain whether the king is compared to the cherub or to a person like Adam, a first man who is associated with the cherub, or how the two are related.[40]

Some have regarded this as another version of the story of Adam's Fall, but this cannot be taken as certain. In my judgment it is more likely to be a Fall: not, however, that of the first man, but rather that of one of the heavenly creatures who were also associated with the mountain of God. This being was perfect; it walked among the precious stones. It was one of the cherubs, but it forsook its work of guardianship, fell into sin and evil, and was expelled. It is suggestive to consider Pope's view that the story derives from an older (Canaanite and Ugaritic) myth of the god El, who was deprived of his pre-eminence and consigned to the underworld. In other words, the story is not so much parallel to that of Adam's disobedience as to that of the 'angelic' fall of 'Lucifer, son of the morning' in Isaiah 14.12 (RSV: 'Day Star, son of Dawn'), who is 'fallen from heaven': he had aspired to ascend to heaven, to set his throne on high, but now he was cast down to Sheol. Ezekiel stresses the 'wisdom' of this being, a feature which makes him similar to the snake of Genesis 2, and dissimilar from the humans of that story, who were conspicuously lacking in that quality. We have here an aspect of the tremendously powerful idea, lacking in Genesis 2–3 itself except for the very limited role there of the snake, of a heavenly or angelic rebellion against God and the casting out of the rebellious heavenly forces.

Ezekiel thus represents an important stage in that process whereby the elements of sin, of blame, of revolt came to be more emphasized.[41] But these elements may have started out as part of the picture of the *heavenly* revolt against God, which in later times was to become so much more powerful. On the other hand the Ezekiel passages, being vague, difficult and textually obscure, were probably a force that in its turn influenced the Genesis story. Indeed one may say that it is upon the combination of this portion of Ezekiel with Genesis that the traditional picture of Adam and Eve as rebels against God depends, whether consciously or unconsciously.

Ezekiel is relevant also for his vision of the valley of dry bones (ch. 37), a passage that shows the existence of the terminology and ideas of bodily resurrection, but one that does not necessarily state the expectation of it. For the passage is an image of the renewal *of the nation*; cf. the image of the joining of two sticks which follows (Ezekiel 37.15–23). Since the revivification of the bones is a symbol, there is no indication that actual bodily resurrection is to follow, and no interest in the resurrection of individuals. The same may be said of other passages like Isaiah 26.13–19; Hosea 6.1–3.[42] Later interpretation, on the other hand, may have seen these references more literally.

8. *The impact of 'sceptical' Wisdom.* The Wisdom tradition is important in this, because, unlike psalms and hymns which might voice vague aspirations, Wisdom had a certain need to work things out intellectually. And some of the major passages I have quoted come from poems that have a Wisdom tinge about them. But the passage of our theme through the Wisdom domain was deeply affected by the more 'sceptical' turn that Wisdom took in post-exilic times, notably in Job and Qoheleth. The effect of this tendency was to question some of the principles that had, in the past, seemed to take care of the problem of death. With Job, it came to be questioned whether righteousness was necessarily rewarded. More seriously, Qoheleth questioned some of the other pillars of conventional wisdom. As he sees it, there is no difference between a righteous man and a sinner: for both are forgotten. In other words, the principle of survival of the name is cast in doubt. Similarly, survival through offspring cannot be depended on. If a man begets 100 children and lives many years, so that the days of his years are many (6.3), but he does not enjoy life's good things, and also has no burial, I say that an untimely birth is better off than he. The fate of the righteous and that of the sinner are the same.

The effect of this argumentation is that the question of continuance, through name and offspring, has to be passed over from normal human remembrance to remembrance by God himself: speaking of eunuchs, who have no children, God insists that *he* will provide a monument and a name, better than sons and daughters (Isaiah 56.4–5). This line of thought is continued by Wisdom 4.1:

> Better than this is childlessness with virtue
> for in the memory of virtue is immortality,
> because it is known both by God and by men ...

Thus some of the main assurances which the older Hebrew belief may have provided are, by this time, becoming more uncertain, through internal developments and inner-biblical exegetical-critical movements.

In the later literature the interests of the themes of sin, blame and death become more prominent.[43] In Hellenistic literature guilt, sex and death were handled in accentuated tones, as compared with classical Greek expressions; consider the contrast of Apollonius Rhodius with Homer. Among Jewish sources, similarly, the story of the angel marriages, still relatively muted and ambiguous in Genesis 6, comes to be enormously expanded in scope and detail. The rebel heavenly beings, the עירים, ἐγρήγοροι or 'Watchers', are embellished with names, with details, with accounts of the fearful proliferations of evil that they introduced upon the earth.[44] At the same time Enoch, a person from the same period, is the subject of the most intense interest, far outweighing the very limited information given to this person in Genesis itself. Genesis itself, we may say, is on the threshold of this development and has to be read in the light of it, but has still not reached the stage of making much of it explicit. In this literature of post-biblical Judaism there is something more of an anticipation of the later Christian ideas of 'original sin' than is to be found in the Talmudic Judaism that we know. And Enoch, as we have seen, is a foremost biblical example for the thought of immortality.

Characteristic of this literature, and fateful for the future, was another development: if guilt and blame were to be so important, they were to be differently apportioned as between man and woman. In Genesis it is true that the woman engages in the first discussion with the snake, but the text gives little impression of putting the blame upon the woman or of worrying at all about the degree of blame attaching to either one of them. This, incidentally, is another reason why we should not see the origins of sin and guilt as the central concern of the story in Genesis. Genesis is not precise about guilt and blame, because that is not a main concern there. In the later literature this is no longer so. It is probable that the cultural changes of the Hellenistic period brought this about. It becomes more important to know whether men or women are to blame. Thus Ben Sira in the early second century BC worries at considerable length about the loss of sleep fathers suffer over their daughters (the unhappiness of the daughters does not bother him at all), ending up with the immortal words:

> For from garments comes the moth,
> and from a woman comes woman's wickedness.
> Better is the wickedness of a man
> than a woman who does good ... (Sira 42.9–14).

or again:

> any wickedness, but not the wickedness of a wife! (25.13).

It was women, after all, who got mixed up with the Watchers and through whose unions wickedness came to dominate the earth in the generation of the Flood. In the Testament of Reuben that patriarch warns his sons against women:

> Women are wicked, my children: because they have no power or strength as against men, they use wiles and tricks in order to draw them to themselves ... It was thus that they allured the Watchers before the flood (Testament of Reuben, 5.1,6).

And in continuity with this is the remark of Paul that the serpent deceived Eve (rather than Adam: II Corinthians 11.3), and this is amplified in a more seriously discriminatory sense by himself or another, in I Timothy 2.14, to the effect that 'Adam was not deceived, but the woman was deceived and became a transgressor'. As compared with the Genesis account itself, which has the woman take the prime place in the action but in no way accentuates any element of blame against her rather than the man, the Hellenistic age was more discriminatory and more ready to blame the woman. Paul fell in with this tendency. Adam, as we shall see, retained a good deal of glory and splendour in spite of his faults, and Eve got rather more of the blame.

9. *Variety of thought on our subject.* This same 'intertestamental' period saw an efflorescence of interest in the questions of immortality and resurrection. As has been said, the first clear affirmation of individual resurrection in the canonical Old Testament is in Daniel,

and even there it is a vague, muted and qualified affirmation. And Daniel was only one in a large variety of texts issuing from about the same time: roughly, the second century BC. All this has been well surveyed by Nickelsburg and I do not have to repeat the ground covered by him. If there are texts that are interested in a resurrection of the body, there are others, like IV Maccabees, that are interested in the immortality of the soul, and neither of them gives any impression of sensing an opposition to the other. There are also texts like I Enoch 103–4 which speak of a future resurrection of the *spirits* of the righteous.[45] And one reason why this is so is that the question is still not defined expressly as one of what happens to individual people *after death*. As I have suggested, one of the main sources of interest in human immortality was not concerned with what happened after death, whether resurrection or immortality, but in the continuation, and perhaps the quality, of life over a long time. Since God was eternal, communion with God must be communion with his eternal qualities. Wisdom 2.23 interpreted the image of God in humanity as implying a sharing exactly in this, in his eternity. In the Qumran scrolls, as Nickelsburg points out, there is considerable celebration of 'eternal' qualities, whether eternal life to be enjoyed by the saints, or eternal pain and derision to be suffered by the evildoers, but little is said to indicate how this is related to *death*. 'Eternal' qualities are qualities that can be gained or merited, or inherited by predestination, and enjoyed or suffered *now*. All this is a background for those later Christian traditions to the effect that eternal life is defined by its qualities rather than by its temporal situation, whether in the future or after death. Yet eternal life in the sense of quality of life, enjoyed here and now, an option that many writers favour for today, is an option that leaves obvious problems within it. Many formulations of it leave it vague and unsettled whether this eternal life continues after death. One can decide not to bother about the question, which seems to have been a common course. But, if the question is asked, it is likely to mean either that this kind of eternal life ceases with death, when the human person dies, or else that because of its eternal quality which is shared with God, it goes on after death – in which case it is more or less bound to mean immortality of the soul after all.

In a memorable sentence already quoted, Stendahl wrote that 'The whole world that comes to us, through the Bible, OT and NT, is not interested in the immortality of the soul.' Now this *might* possibly be argued if one restricts one's view entirely to the canonical books and says absolutely nothing about the 'world' out of which they come. But to say that it is true of the 'whole world' that comes to us through the Bible is sheerly impossible. The world of late Judaism out of which the New Testament came is burning with speculation about various options, and the immortality of the soul is one of the favourite candidates. How was it possible for Stendahl, in saying this, to express appreciation for what he has learned from Nickelsburg's book,[46] when that same book is full of the evidence for this variety, including ample evidence for belief in immortality of the soul? The weakness of his position is betrayed when he speaks of the biblical scholar as one who 'specializes in sixty-six so-called books that do not know of immortality of the soul'. Even if this should be true of the sixty-six books, the argument is valid only if the narrower Protestant canon is taken in a very exclusive way. The moment one considers the wider canon, including Wisdom and Maccabees, the argument is destroyed. And since in the last resort the argument is working towards what Paul may have thought or known, its invalidity is total: for in at least some of the relevant areas, we have seen that his really close congeners lie in the territory between the books of the Hebrew Bible and the beginnings of Christianity. The importance of this will be further reinforced in the pages that follow.

10. *Martyrdom.* One of the forces which sharpened all these issues, and which weighed heavily in the scales in favour of immortality of the soul, was the matter of *martyrdom.* Religious martyrdom, as it emerged in the Maccabean period, was something of a new thing in the history of the world. There had always been killings and massacres of people because they were enemies, foreigners or otherwise disagreeable, but the Maccabean period was perhaps novel in that physical force and continual torture were used *precisely in order to enforce conformity* to a religious or ideological order. One could escape from this ghastly suffering simply by saying certain simple formulas or undertaking some simple acts. If one did not conform to these demands, the body would be gruesomely tortured and finally destroyed.

In this situation talk about the Hebrew conception of the totality of the person or about the refusal to accept any split between body and soul was of little use. It is highly probable that the necessary vocabulary for the distinction of body and soul, or body and spirit, was available by this time in the relevant Aramaic or Hebrew. The body could be physically crushed and destroyed by the oppressor. A future resurrection could, indeed, promise a restoration of that body. But, on the other hand, one of the major solaces of the persecuted was the claim that, in spite of the shattering of the body, he or she had maintained steadfastly the religious loyalties *of the soul*, and precisely that fact made the hope for a future resurrection more meaningful and more sure. Through the endurance of suffering one may show forth the control of the flesh and the bodily passions, a control which is a virtue in itself, as IV Maccabees makes clear.

Conversely, the immortality of the soul made it all the more essential for the persecuted to remain steadfast, for with it in mind they could not easily comfort themselves with the thought that it would all be over, and be forgotten, when death finally came. No, for even after death it would be found that the soul had survived and would have to pay by yet further sufferings for any failure to resist the earlier sufferings of the body. Conversely, the persecutors would in due course find that *their* bodies, now so strong, would waste away, but that they, in soul or in body, would have to enter into major sufferings such as the martyrs had endured. Moreover, yet again, the argument from immortality had this special power: that immortality took effect *at once*. Resurrection might no doubt come along in due course, but one would have to wait, for it was beyond doubt that in the short term dead persons were in a bodily sense dead. If there were rewards for a good life and for a martyr's testimony to be gained, immortality provided a path through which one could gain immediate access to them. In this situation, then, immortality of the soul and resurrection of the body worked together. And this, we may add, is very similar to the way in which the symbiosis of the two concepts operated in traditional Christianity.

11. *The Wisdom of Solomon*. Many of these different threads were brought together by the Wisdom of Solomon. This book unquestionably adopts into a basically Jewish frame of reference certain ideas derived from Greek philosophy. Like many such Jewish works, it did not necessarily accept, and did not necessarily even know, the argumentative linkages upon which these ideas were founded by their Hellenic authors. On the other hand, we have already seen how it was a central link between the older Hebrew traditions and the thought and interpretations of St Paul. Particularly important to it is the theme of immortality. In it, man was

created with, or for, immortality, as we saw, and it was through the devil's envy that death came into the world. Righteousness is immortal (1.15). Wisdom is the source and fountain of immortality, and makes it available and accessible to all. However, above all, in the area most important to our present subject, Wisdom opts decisively for the immortality of the soul, but without thereby necessarily questioning the future resurrection of the body. In ch. 3 the author reasons against that sceptical tradition which pointed out how the name would be forgotten. He points out how, if that tradition is valid, 'there is no return from our death' (2.5), how in consequence there are no ethical sanctions and the impression is left that one may as well act unrighteously and violently. But, he concludes, all this is wickedness, because 'they did not discern the prize for blameless souls (οὐδὲ ἔκριναν γέρας ψυχῶν ἀμώμων)' (2.22). The hope of the ungodly man is like chaff carried before the wind (5.14), but 'the righteous live for ever, and their reward is with the Lord' (5.15). Thus the contrast of body and soul, so much abominated by modern theology, is clearly embraced, and with it the idea that the body, being perishable, is the cause that damages the powers of human reasoning:

> For the reasoning of mortals is worthless
> and our designs are likely to fail
> for a perishable body weighs down the soul
> and this earthly tent burdens the thoughtful mind (9.14f.).

Thus, coming at the end of the Wisdom tradition of the Old Testament, and forming as it does an all-important link with Paul's theology, Wisdom brings together themes inherited from the earlier stages of the Hebrew Bible and embraces certain themes of Greek philosophy as a uniting force that brings them all together and provides a synthesis and an answer.

> The souls of the righteous are in the hand of God
> and no torment will ever touch them.
> In the eyes of the foolish they *seemed to have died*,
> and their departure was thought to be an affliction …
> but they are at peace.
> For though in the sight of men they were punished,
> their hope is full of immortality (Wisdom 3. 1–4).

Indeed, this is doubtless a dangerously innovative suggestion,[47] but one can imagine a scenario under which the immortality of the soul was a received option in Jewish thought *earlier than*, and provided a catalyst for, the formulation of the resurrection of the body. After all, definite formulations of the resurrection are strikingly late: the Sadducees are reported to have denied it, and resurrection was still not a universally received opinion, as St Paul was able to turn to his own advantage through a dispute between them and the Pharisees (Acts 23.6–8). Immortality of the soul was surely already known on the Jewish scene before Daniel 12 provided what is supposed to be the first, and very weak, formulation of bodily resurrection in canonical scripture. Given that immortality of the soul was already known in the times of persecution in the earlier second century BC, one could readily imagine how it was thought: Yes, my soul will survive after death, but what about my body which was broken by violence? Will not the persecutors have won if it is left unrepaired? Will it not be proper recompense for me that my broken body be restored and rejoin my eternal spirit? – Yes, indeed, the resurrection that is to come will do just these things. In other words, the conviction of immortality could have led to the need for resurrection. Such a dialogue is not, perhaps, to be proved, but something of the sort remains possible and makes sense of some of the difficulties.

12. *Conclusion*. I do not suppose that in this chapter I have proved the immortality of the soul; it was not my purpose to do so. Nor has it been my purpose to argue that the immortality of the soul is a good thing, or a bad thing, for people to believe in. What I think I have shown is that, for much of the Hebrew Bible, death, so long as it was in proper time and in good circumstances, was both natural and proper in God's eyes; that the Old Testament provides thoughts and aspects out of which ideas of the immortality of the soul could naturally and easily develop; and that the world on the basis of which much of the New Testament was written was a world in which the belief in that immortality was lively and strongly represented.

3

Knowledge, Sexuality and Immortality

In the first chapter I sketched out the main outlines of a position about the story of Adam and Eve, and argued that the main emphasis of that story fell not on the origin of sin or evil, and not upon the entry of death into a world where there had been no death, but upon the loss of a chance of immortality which might conceivably have been seized by humanity. But I did not go into every detail of the story; certain aspects remain to be filled up, and of these the most important is the question of the tree of life.

Now it must be admitted that the role of the tree of life within the Genesis story has difficulties and apparent inconsistencies. How were the two trees related, the tree of knowledge and the tree of life? The text leaves us confused about this. In Genesis 2.9 it tells us that God made all sorts of trees to grow, 'and the tree of life in the midst of the garden and the tree of the knowledge of good and evil'.[1] But in v. 17 the only tree that is forbidden is the tree of knowledge of good and evil, and nothing at all is said about the tree of life. In 3.3, when the woman is explaining to the snake which tree is forbidden, she defines it as 'the tree which is in the middle of the garden', and this is the only tree that is forbidden. That this was the tree of knowledge of good and evil follows naturally from what the snake says. It also follows from the words of God in 3.22, which accept that the humans have 'become like one of us, knowing good and evil'. It now becomes important that they should not eat of the tree of life, which in fact has not been mentioned at all since the trees were first mentioned at 2.9. What is not clear is: why the tree of life was not forbidden as the tree of knowledge was; where it was in relation to the latter tree; whether the humans had in fact been eating of the tree of life all along; and, if they had not been eating of it, why they had not done so.

One of these questions, however, can be quickly answered, and provides a good datum point for our discussion. We begin by recognizing one basic fact: that the phrase of 3.22b, 'and now, lest he put forth his hand and take also of the tree of life and eat and live for ever', must mean that the fruit of the tree of life had not been previously eaten. By Hebrew grammar and meanings, I believe it cannot mean 'lest he continue to eat of the tree of life, as he has been doing all along'.[2] In other words, the phrase which states the motivation for the expulsion of the humans from the garden is a phrase that excludes the idea that they had been eating of the tree of life from the beginning. The possibility that they might eat of the tree of life is something new, a threat of a radical change in the human situation. Admittedly, there is no indication of why they had not eaten of this tree, or of why the eating of it was not prohibited as the eating of the other tree was prohibited, or indeed of how it was related to the tree of knowledge at all. But for some of these questions an explanation can be offered, and is supported by many scholars in the field, as follows.

In its older stages, according to this explanation, the main story, as its own structure shows, concerned only one tree. Only at the margins is the tree of life mentioned at all: first when the garden is first planted, and secondly at the end of the story, when the human pair is to be expelled from the garden. Originally, seen in this way, the tree of life belonged to a different story and a different theme. This explains why the story as we have it says nothing about the *relation* between the two trees, nothing about the position of the tree of life in relation to the other, and nothing about forbidding access to the tree of life. The basic story was about one tree, 'the' tree, a tree which, it is suggested, did not have any name or designation, and which came to be called 'the tree of knowledge of good and evil' because that was its function in the development of the story. It was when the other tree was added, and was expressly named as 'the tree of life' (because that was essential to its function), that the original (nameless) 'tree' received its designation too, as 'the tree of knowledge of good and evil'.

Thus, if this explanation is taken seriously, the story at an older stage was about 'the tree', the one tree which was forbidden. The woman and the man disobey the prohibition and are punished by expulsion from the garden, not because they might find access to the tree of life (which at that stage was not there at all!), but simply as a punishment or deprivation. They had been put in a lovely garden, now they were expelled. They did not do what God wanted, and so he threw them out. 3.23 expresses this: he sent the man forth from the garden of Eden, to till the ground from which he was taken (i.e. to fulfil the punishment indicated in vv. 17–19). That was the end of the story, for there was nothing there, at that stage, about the tree of life, since vv. 22, 24 are part of a different story, added later.

Now this explanation may well be, historically, correct; the story may well have developed in just such a way, and the interpretation fits with many of the details. It is simpler if it is so taken. But if we go no farther than that, the effect is to marginalize completely the theme of the tree of life. Even if the story grew up in the way I have described, in the story as we have it the tree of life is very obviously important. Even if it comes from a different story or legend, it must have been important to the person or tradition who introduced it, and that person or tradition not only introduced it but made it into the sole express motivation for the expulsion from the garden. Therefore, even if the inclusion of the tree of life leaves a number of loose ends and discrepancies, it cannot but be important, for it alters very substantially the total general direction of the story.

My approach, then, is one that gives full value to the element of the tree of life, since it is important for our present text. We are thus following present fashion and taking a 'canonical' approach, giving full value to the 'final text'. On the other hand, I consider that an approach through the 'final text' does not negate, but positively demands and benefits from, an attention to the previous stages out of which the final text has been derived. It is interesting to note, in any case, that in this particular instance the canonical approach, focusing on the final text, results in a more surprising and untraditional interpretation than canonical approaches are commonly expected to do.

Or we may restate this aspect as follows: even if there was once a separate story which concerned the one tree (of knowledge), that story even in itself cannot have concerned a human pair who were immortal before their disobedience and who were only thereafter reduced to mortality. On the contrary, it concerned mortals who had been comfortably placed in the garden but were now faced with a much more miserable life. One can thus understand how such a story could be integrated with another story about a tree of life which could bring immortality and which came near to being enjoyed.

Let us put it in another way. There may have been two stories, which came in the end to be interwoven. One was about a single tree, the eating of which was forbidden, but which brought with it as consequence the knowledge of good and evil as well as expulsion from the garden. The other story was about the tree of life, which would normally be remote but on this occasion became practically accessible to humans, but nevertheless the chance of tasting it was lost to them. The two are mixed up in a somewhat vague and unclear manner. But there is certainly a meaning and a message in the fact of this mixture, ill-coordinated as the mixture at some points may be. In the story as we have it, the tree of life is nearby at the beginning, and in the end it forms not perhaps the only but certainly the most vividly expressed motivation for the expulsion. Even, therefore, if the entire element of the tree of life was absent from the older form of the story, the person or tradition who brought in the tree of life must have had a purpose in doing so, and an awareness that by introducing this element he was substantially altering the meaning of the story, or creating a new story the general purport of which was a very different one. It therefore seems to me impossible simply to leave the traditions of the tree of life out of our exposition.

That there may well be a connection between knowledge and immortality is clearly indicated by the story of the Garden of Eden itself, for the two trees that form the centre of its action are the tree of knowledge and the tree of life. Undoubtedly these two are not identical, for their very names and their parts in the story expressly distinguish them. In spite of their difference, however, it is likely that they have something in common. What they have in common is that both, in some degree at least, are prerogatives or attributes of God. 'You will be as gods, knowing good and evil,' said the snake, and God himself admitted that this prediction had come true: 'Behold, the man has become like one of us, knowing good and evil' (Genesis 3.22). If humanity had come this close to deity, it was only one further step for it to partake also of the tree of life, and, it is implied, if immortality were added to knowledge, the approach of humanity to the status of deity would become intolerably close. In the event, therefore, humanity is left with knowledge, but immortality is denied to it. Man has come close to God in the one regard, but not in the other.

That the tree of life has a connection with wisdom and knowledge is confirmed by the fact that elsewhere in the Old Testament where the phrase 'tree of life' occurs, it is in the 'wisdom' book of Proverbs that we find it. 'The fruit of the righteous is a tree of life' (Proverbs 11.30); but, most of all, wisdom herself is a tree of life: 'she is a tree of life to those who lay hold of her' (Proverbs 3.18); cf. also 13.12, where a desire fulfilled is a tree of life, and 15.4, where a gentle tongue is a tree of life. These usages give us some impression of the suggestions conveyed in Israel by the expression 'tree of life'. It is beyond doubt, indeed, that the idea of such a plant was inherited by Israel from much older folklore or mythology, of which the best and most pertinent example is the plant of life in the Gilgamesh Epic.[3] If, in its older origins, and still in Genesis, the tree of life was above all a symbol of eternal life, in Israel it came also to be a central symbol for the 'wise' and prudent ways of conduct within which one ought to live.

What then of the phrase 'good and evil'? The tree of knowledge is characterized several times in precisely this way, as the tree of knowledge of good and evil, and that good and evil are the products of its fruit is made plain not only by the declarations of the snake beforehand but also by those of the deity himself afterwards (Genesis 3.22). But what is meant by 'good and evil' in this context? Scholars have offered a surprising variety of different explanations of these apparently simple words; we cannot go out of our way to discuss in detail the virtues and faults of each one.[4] A few, however, must be mentioned, because they will have a connection with other topics which we shall wish to discuss. One view, which is Westermann's own and likely to be well known, insists on the *functional* nature of 'good and bad'. The fruit of the tree does not impart grand principles of morality, but knowledge of what is useful or injurious, practical matters like choosing good foods rather than bad, good ways of finding or transporting water rather than bad ones. Going in another direction, some scholars have emphasized the *sexual* character of this 'good and bad'; it is the consciousness of sex: 'begetting and sparking life is a divine craft' (Gressmann). According to this point of view, a good example is found in Barzillai the Gileadite in II Samuel 19.35f., an old man coming to greet David after the Absalom rebellion, who uses the same phrase, and says, 'I am now eighty years old; can I discern between good and evil (RSV what is pleasant and what is not)? Can your servant taste what he eats or what he drinks? Can I still listen to the voice of singing men and singing women?' – in other words, at his great age he has lost the perception of pleasure, in sex, in food and drink, in music. If the sexual interpretation is a narrowly particular one, the opposite is a highly generalized one: 'good and evil', taken in this way, is a stock phrase like 'high and low', 'lock, stock and barrel'; you will know everything. 'The knowledge of good and evil,' says von Rad, 'means a knowledge of everything in the broadest sense of the word.'

These, then, are some, though not all, among the current scholarly explanations. For myself, I must say that I do not find any of them very satisfying, for various reasons which will become clearer as we discuss other aspects of the story. To me it seems most likely that the power of rational and especially ethical discrimination is meant. This is something that belongs pre-eminently to deity, and particularly so in Israel. Before their disobedience the humans had no need for any such power. Their disobedience gives to them that power, but with it the perception of their own weaknesses and limitations. I do not find it clear why the 'functional' knowledge of what is useful or harmful, important as such knowledge is, should count as making human beings 'like God'; and the same objection stands against the sexual explanation. To say that it is 'knowing everything' seems to be the abandonment of an explanation: I do not see how Adam and Eve knew *everything* at that stage of their career, or ever. Of the various biblical passages that use the phrase טוב ורע, some seem to me not to be a parallel except under misleading semantic arguments, and the one that strikes me as closest is II Samuel 14.17, where the woman of Tekoa, certainly a 'wise woman' in the understanding of the time, tells David that 'my lord the king is like the angel of the Lord to discern (more literally: לשמע הטוב והרע, to hear, to perceive by hearing [the difference between]) good and evil'.

Knowing the difference is the essential thing. Connected with this is the matter of the *nakedness* of the human pair,[5] an aspect certainly highlighted in the text. What is the meaning of the stress on this point? Before their disobedience they were naked and 'were not ashamed', as traditional translations put it. What is the cultural background of this? In Hebrew culture clothing was fundamental for normality. In the beginning the first humans were naked and were not disturbed, not embarrassed, by this; they were like the animals (for, remember, Adam was with the animals, was familiar with them, and gave them their names. 'That's a horse,' he said, 'That's a cow.') Eve was created as a sort of super-animal, because the existing animals did not fulfil his needs. No consciousness of distinctions was necessary, no one gave the matter a thought. The arrival of 'knowledge' brought discrimination. Things differed according to various categories. Human was different from animal, and, as within the animal world, so within humanity, there were fences between categories.

Much traditional Christianity, understanding the story as a serious rebellion against God, a sin the consciousness of which would obviously be guilt, thought of the awareness of nakedness as a manifestation of guilt, expressed as shame. One is ashamed for the evil one has done, and this shame for evil done is expressed as a feeling of nakedness. But this is not the meaning of the Hebrew words. התבשש, conventionally translated as 'be ashamed', often means rather 'be embarrassed', 'be shy', 'have a sense of being let down', 'be disappointed'. Adam and Eve hid among the bushes because they were embarrassed to be naked; they did not want God (or anyone else) to see them like that: it was this, rather than shame at an evil deed, that motivated them. It was not shame caused by wrong-doing, it was rather a matter of propriety. It is true that terms that may also mean 'cover' are found in the vocabulary of atonement for sin, but these terms, even if they do mean 'cover', are not terms used of clothing, which is the essential point here. Hebrew culture was built upon clothedness as normality. 'Clothing, as something which is part of this life, belongs to human existence in the world in which people live.'[6] Strict prohibitions of the 'uncovering of the nakedness' of others within the family group were familiar. Nakedness could exist, but only under special conditions if propriety was not to be seriously violated. Between married persons it could be taken for granted, and there was nothing against that, though it is interesting that little is positively said, anywhere, that celebrates nakedness as a positive source of pleasure: even the Song of Songs, though richly celebrating physical beauty, has a certain reserve and a tendency to stick to metaphor. Within other family relations nudity was strictly forbidden, with a strictness that only goes to prove how much it needed to be forbidden. Occasional transgressions of these rules, as with Noah himself and with Lot (Genesis 9.20ff.; 19.30ff.), only went to emphasize the matter through the importance of the persons involved. Prostitutes, we may be sure, normally kept themselves clothed: we know this from the story of Tamar in Genesis 38, which tells how she, having removed her widow's garments, 'put on a veil, wrapping herself up', so that her father-in-law Judah, seeing that she had covered her face, thought it clear that she was a prostitute, and thus, it follows, is explained the fact that he did not recognize her and had no idea who she was – until later.

Certain exceptional circumstances made nakedness quite proper, but these circumstances were rare. One such lay in prophecy, well-known as an area of eccentric behaviour. Isaiah himself is said to have walked about naked and barefoot for three years, as a prophetic sign (Isaiah 20). Similarly, in the earliest days of prophetic history in Israel, in the time of Samuel and Saul, the manifestations of prophecy included ecstatic experiences with music, with dancing, and with trance-like lying naked. Saul sent a succession of messengers to capture David when he was with Samuel and other prophets, and all the messengers were themselves infected by the prophetic spirit; finally Saul himself went there, and he too was seized by that spirit, stripped off his clothes and lay there naked for twenty-four hours (1 Samuel 19.24). No wonder that they asked the question, 'Is Saul also among the prophets?' But, except for such special occasions, nakedness was a matter of deep embarrassment and disgrace. David at a key point in the story of his life sent an embassy to Hanun king of the Ammonites, and that king seized the heralds, shaved off half the beard of each, cut off their garments at the hips, exposing their sexual parts, and sent them away. The men were deeply ashamed, נכלמים מאד, not the same word as we have in Genesis but more or less the same idea (II Samuel 10.5). It was one of David's strokes of genius that he went personally to meet the men when they were thus disgraced. This was no matter of moral guilt, but of disgrace at a gross contravention of propriety. David himself got into some trouble when dancing before the ark as it was brought into Jerusalem (II Samuel 6), Michal his wife, daughter of Saul, objecting to his inadequate dress: we may perhaps surmise that David thought himself to be in that special category of prophetic experience which I have just mentioned. However, Michal was more likely expressing the general view of her own culture and that of the Old Testament as a whole.

Where does this lead us in respect of our general subject? If I am right, it confirms that the matter of nakedness and 'shame' is not primarily a matter of shame for a sin committed. Certainly there was a disobedience, and the human pair were well aware that God would be annoyed, as he was; fear of divine anger and the sense of nakedness doubtless complemented one another. But it was the awareness of nakedness that counted as the effect of the new human situation. It was not primarily an awareness of past disobedience to God, but rather a coming of consciousness of lines that must not be crossed, of rules that must be obeyed, and in this sense a discernment of 'good and evil'. It was certainly also a coming of self-consciousness: as God pointed out, how else did Adam and Eve know that they were naked? If I am right, this understanding of the matter of nakedness confirms the explanation I have given of the knowledge of 'good and evil'.

This leads us on to another aspect, which is more uncertain but none the less fascinating: the question of the original relation between Eve and the snake. The clue to this lies in the name Eve itself. Genesis provides a folk-etymology, as with many names in the book: 'the man called his wife's name Eve, because she was the mother of all living' (Genesis 3.20). This makes the obvious connection with the common lexeme חיה, 'live, life'. But the etymological explanations of biblical names are, obviously, mostly wrong.[7] Though many or most Hebrew names that were in actual use were semantically transparent and carried a meaning, most of those that were thus transparent were not explained in the biblical text, and the ones for which explanations were furnished were mostly those of which the meanings were unknown. The explanations given are therefore folk-etymologies based on associations of some kind with words or aspects of the text in which the names appear. In fact, to come to the point, it has long been suggested, though never clearly proved or evaluated, that the name Eve was related to an Aramaic חויה and similar Arabic cognates, which mean 'snake', and that the mythological ancestor of Eve was some sort of 'serpent goddess' who was, perhaps, the goddess of life.[8] The point has never been proved and is perhaps incapable of proof, and yet it has much that is attractive about it. For it fits in with some things that are otherwise peculiar in our story. Why, after all, a snake? And why in the crucial decision to eat of the fruit of the tree does everything stated come from the woman, while the man says nothing and does nothing, except to take the fruit and eat it after she has done so? In the story as we have it, indeed, she is a definitely human female, but the proportions and structure of the story make more sense if at some earlier stage she was something more than that. Had she perhaps at some time been a snake-goddess of knowledge and wisdom who tempted the first man, pointing out to him the excellent qualities of the fruit of the forbidden tree? Though she was later a human female, this background might explain why the human female takes so leading a role.

The aspect of sexuality has another side to it. As I have mentioned, some have interpreted the course of the story as the achievement of awareness of sexuality.[9] The awakening of the sense of sex, the difference between the sexes, and so on – these explanations have been invoked by scholars such as Gressmann, H. Schmidt and Gunkel. I must confess that my own opinion is the reverse. The question rests, however, on several points which are not made absolutely explicit in the text and depend rather on our estimate of cultural probabilities.

Two great points are not made explicit in the text: (*a*) how long were they in the Garden of Eden before their disobedience?, and (*b*) did they or did they not make love during that time? The two questions are interconnected. The chronological question affects the sexual one. How long were they in the garden before the snake came along and initiated his fateful conversation with Eve? Half an hour?[10] A day? A week? A year or two? Much longer (according to Jubilees 3.17, they were more than seven years in the garden before the snake intervened)? Much appears to depend on this. For if they were there for a year or two, and if they made love during that time, one would have expected that Eve, a fertile lady, would have produced her first child. Yet no child is reported before the expulsion. The first sexual intercourse actually reported comes after the expulsion, when we hear that Adam 'knew' his wife and she produced her children, Cain and Abel. Of course, from the modern critical point of view, one can say that that is a different story, and so it is; but, even if we discount this, and insist on taking the text as a whole, it does not clearly imply that they had had no sexual intercourse before this time.

In my judgment it is far more natural to understand that the human pair did make love in the Garden of Eden. The effect of their disobedience was not that they became aware of their sexuality, but the reverse, that they became aware of the negative aspects and limitations that were henceforth to accompany it. We have here the same question of cultural assumptions as we saw in the case of death: just as the natural cultural assumption was that humans were innately mortal, so it was the cultural assumption that they were sexually aware and active. That they were 'naked and not embarrassed' fits in exactly with this. It was as in the world of animals: sex was obvious and immediate, and they did not worry about it at all. It was like that ideal sex world perceived in the Song of Songs but seldom practically realized. The acceptance of sexuality is normal in Hebrew culture, and it is only from the profound sexual hang-ups of Christianity that could have come the idea that there was no sexual activity 'before the Fall' (to revert, for a moment, to the customary phrase).[11] Thus Bloom correctly writes: 'What is it to know good and bad? Again the normative misreading has reduced this issue to the knowledge or consciousness of sexuality, but J has too healthy a view of human sexuality for such a reduction to be relevant or interesting.'[12]

But perhaps the most powerful reason is this: in biblical Hebrew narratives, not only in this first story but throughout, the natural assumption of sexuality is demonstrated by the fact that there is no curiosity, even no positive interest, in it for its own sake. No one has enough interest to write a description of how good it was to make love (poetry, with the Song of Songs, could make an exception here). *All* narratives of sexual intercourse are there for family reasons: the birth of children, sometimes competition between wives and concubines, sometimes the safeguarding of a family line, sometimes an adultery which will produce troubles and further narratives, things like that. No one wants to hear about sexual relations for themselves, such as how many times persons made love, or in what ways, or whether they were clothed or naked, or whether it was always equally good. The intercourse which leads to the birth of Abel is the first one reported because it is the first one that leads to the birth of a child.

Milton was obviously right in seeing that they made love (*Paradise Lost*, *IV*, 741ff.):

> nor turned I weene
> Adam from his fair Spouse, nor Eve the Rites
> Mysterious of connubial Love refused ...

and Milton went on to excoriate the 'hypocrisy' of those who supposed it would have been better if no such thing had taken place:

> Whatever Hypocrites austerely talk
> Of puritie and place and innocence,
> Defaming as impure what God declares
> Pure, and commands to some, leaves free to all.
> Our Maker bids increase, who bids abstain
> But our Destroyer, foe to God and Man?

Several further considerations can be added to support this view:

First, it is true that no sexual intercourse is reported by Genesis until after the expulsion, but equally there is no express mention that there was no such intercourse. If it had been significant for the text that there was no intercourse, then it would have been natural for the text to say so; but it did not. The cultural assumptions would be that, in the circumstances, the couple would make love, especially in view of the statement made that they were naked. No one has suggested reasons, within ancient Hebrew culture (unlike, say, mediaeval Christian culture), why it would be considered desirable for the couple to have abstained from making love while in the garden, or why there would be point in such an abstention. The alternative would be that temptation and disobedience followed so swiftly that there was no time. But again the lack of interest in temporal exactitude on the part of the text (on this question, much contrasting with some others) makes this implausible. They had no idea that the time in the garden was so short as to eliminate the making of love.

Secondly, it has often been argued that the statement after the expulsion from the garden, 'and Adam knew Eve his wife, and she conceived and bore Cain', intrinsically suggests that this was their first act of love (Genesis 4.1). This, however, is not so. Practically the same words are used of Elkanah, the father of Samuel: 'and Elkanah knew Hannah his wife, and the Lord remembered her; and in due time Hannah conceived and bore a son' (I Samuel 1.19). But Elkanah and Hannah had been making love for years: the trouble was that she had produced no child. Genesis 4.1 is there, not because it is the first human act of sex, but because its function is to introduce the first son of the original pair.

Thirdly, Jubilees, as we have seen, understood the time in the garden to be quite long, over seven years, and even after the expulsion allowed another long period before their sexual activity began: 'And they had no son till the first jubilee [i.e. more than fifty years after creation]; and after this he had intercourse with her' (3.33). After Abel's murder things grew even worse. 'They mourned for Abel four weeks of years [twenty-eight years] and in the fourth year of the fifth week their zest for life returned, and Adam had intercourse with his wife once more, and she bore him a son, and he called him Seth' (4.7). But there is a reason behind this slow movement. Jubilees, a book extremely interested in chronology, was trying to take into account the figures of the genealogy in Genesis 5, according to which Adam was 130 years old when Seth, the son who counts as first for purposes of this genealogy, was born. If it took him 130 years to produce the son who for later genealogical purposes was to count as firstborn, a reasonable number of years before Abel and Cain were born is to be expected also. Jubilees is spreading out the time of this early life, when things moved more slowly.

This, however, was only one extreme among early interpretative schemes. At the other we have Rabbi Eleazar b. Azariah, according to whom, with a useful application of miracle, suitable for the exceptional situation, things were accelerated as much as in Jubilees they were retarded, and all these events took place within one single day, three miracles all in one day: on that one day they were created, they made love, and they produced offspring.[13] So it was all temporally very concentrated. In fact the chronology of Adamic sex is pleasant and amusing, but ultimately irrelevant: clearly the text had no thought of the question and gives no indication about it. For a time they were in the garden, and later they were expelled. Milton's time-frame is as good a guess as any, and very properly gives them time to enjoy one another adequately before disaster struck them.

Fourthly, another point will not have escaped the attentive reader: given the striking announcement, immediately after the creation of Eve, that a man is to stick like glue (this is the meaning of the Old English *cleave* here, and of the Hebrew) to his wife, and they to become *one flesh*, it is something of an anticlimax if this first couple did nothing for months or even years to realize it.

To sum up this point, then, it is much easier to suppose that sexual activity was taken for granted in Genesis 2–3 but for that very reason was not explicitly mentioned.

Sexuality, then, was assumed from the beginning, as was mortality; disobedience brought difficulties into it, first of all through greater self-consciousness, secondly through taboos and limitations, and thirdly through God's reaction against their disobedience, expressed especially in the words of disapproval addressed to the woman, whose desire is to be to her husband but he is to have dominance over her.

Somewhere in these relations lies another point: the clear play on words between the *cunning* (ערום) nature of the snake, and the word 'naked' (עירם) applied to the human pair. That this is a play on words is clear, but it is not so clear what its message is. Perhaps we could say: the cunning of the snake offers the humans a source of knowledge, but the knowledge that they receive has its first manifestation in the awareness of their own nakedness. This is what it is going to be like to have the knowledge of good and evil: it is real knowledge, it puts the humans at least in part into the same class with the divine, but the first thing they are going to learn from it is trouble for themselves. Innocence is lost. Knowing good and evil may be a privilege, but it is one that reacts unpleasantly upon one's own self-understanding. Thus the clear sexual interests of the passage may be evaluated without our reaching the unlikely conclusion that the dawn of sexual awareness is the central meaning of the story.

Another aspect that calls for comment is the interest in technology or, as it may be called, technogony, the birth of the arts and crafts. Many stories of the beginnings of the world are interested in this: one that is much more complete than the fragmentary biblical pieces is the work of Philo of Byblos, set in Phoenicia, based at least in part on Phoenician traditions, and having many points of contact with Hebrew material.[14] In the Old Testament the most important such narrative is the story of the descendants of Cain in Genesis 4, from whom came the first shepherds, the first metal-workers, the first musicians, also the builder of the first city. A technogonic interest would fit with the general estimate I mentioned earlier, according to which the 'knowledge' of good and evil was the practical knowledge of what was harmful, what useful. But the content of our story tells against this. The tale of the Cainites is a side-issue in the Old Testament and does not lead anywhere much. Unlike the descendants of Cain later on, Adam and Eve do not learn anything technological. Covering themselves with leaves indicated the absence of technological improvement. Clothes, indeed, are now necessary for the humans, but it is God who makes them, not they themselves.[15] It is reasonable, indeed, to think that our story tells of the start of civilization, but if we so term it we should understand that this story, as distinct from others, shows no interest in technological achievement or in political organization: for it, civilization, if it means anything, means the making of distinctions, especially of ethical distinctions, and the consequent burden of differences, limitations and regulations.

Along with this we have to consider how our story relates to that all-important concept from another stratum of Genesis, the concept that humanity is created in the image of God. There is a clear analogy between the two forms of discourse. The image of God involves some sort of likeness, and humanity in the Garden of Eden has come to be in some measure 'like' God. If 'knowing good and evil' is far from clearly defined in our story, neither is the image of God clearly defined, and controversies about its nature have continued throughout the centuries.[16] If knowing good and evil has some sort of connection with immortality, so also the image of God seems to be something lasting, which does not go away. Some traditional Christianity, insistent on the centrality of the Fall, often maintained that the image of God in humanity was seriously defaced, damaged or diminished by human disobedience, but as is well known, none of the texts which speak of the image of God suggest that it was defaced or diminished in any way at all. In Genesis 9.6 murder was to be punished with death because humans had been made (and therefore still were) in the image of God. In fact, the stratum (P) that talked of the image of God had no incident of a 'Fall' (even if there was one in Genesis 3), or at least none at this particular stage (one may have come later, see the next chapter).

One aspect, related both to the 'image of God' concept and to the story of the Garden of Eden, is the relation of humanity to the animals. Both stories take an interest in this theme. In the first chapter of Genesis, a more systematic account and one usually supposed to have been composed later, humanity is created as the last element in a series lasting six/seven days and including eight deeds of creation. The outer framework of the world is created first, then the plants, finally the animals, and lastly humanity, which alone in the set counts as having the image of God. In the story of Genesis 2–3, by contrast, nothing much is said about the outer framework of the world, but first a man is formed, then a garden is planted and the man is put in it, the prohibition of eating from the tree of knowledge is added, and then God observes that it is not good that man should be alone. He forms out of the soil all the animals and brings them to the man to see what he would call them. The man gives names to the animals, but it seems that this is not enough: man himself does not have a companion or opposite number suitable for him, and God then from his side creates the original woman. All this is well known and familiar.

But there are two striking things about it. The first is that God must have been pretty naive to suppose that the first man would have found satisfactory companionship in a lot of cows and sheep, enlivened perhaps by an occasional lion or leopard, and that this would have provided for him the sort of thing that was later, by default, provided through the presence of a woman. The idea that woman was an afterthought is a remarkable conception, and one that was later corrected by Genesis 1, which made it clear, or at least fairly clear, that male and female were created together from the beginning. Of course there are mythological parallels to the idea, but within the context of Israelite culture it seems highly incongruous. The effect of it is to emphasize the distance existing between the man and the animals; man and woman, by contrast, form a closely-knit and united pair.

The complementary aspect to this is the differential talents or faculties assigned to humanity as against the animal world. In Genesis 1 this is marked by the idea of the image of God, which applies only to humanity and not to the animals, and leads on to the idea of the dominion of humanity over the animals and the rest of the natural world. In Genesis 2–3 it lies in the knowledge of good and evil, which belongs to the world of the gods and removes humanity from the ambience of animal behaviour.

This knowledge, then, involves a contact with universal and eternal principles. These may be poorly known, but their existence is implied by the fact that humanity knows 'good and evil' at all. 'Good and evil' are thus, even if vaguely defined, vaguely defined because they belong to the world of the gods, to an eternal realm where humans have no proper or normal access. It was not, therefore, a completely wild or inappropriate exegesis when early theologians saw in this a grasp by the soul of eternal verities, a faculty which, in the tradition of Greek philosophy, indicated a sharp distinction between human and animal. It was not at all absurd when early interpreters, as early as the Wisdom of Solomon, saw in these traditions connections with the language of Greek philosophy. The power of knowledge endows humans with a transcendence over the absolute limitations of their physical existence. The same knowledge, however, brings with it self-knowledge, and in particular self-consciousness plus the possible awareness of fault and shame. Moreover, the fact that man fails to add immortality to his knowledge only reinstates on another level his weaknesses and limitations. Knowledge may involve contact with the eternal, but sickness, mortality and other aspects of the human condition bring about another set of tragic weaknesses.

In this respect we may find that our arguments bring us full circle. I began by casting doubt upon the traditional supposition that the disobedience of Adam and Eve brought death into the world, in the sense that they had been created immortal and now had to die. On the contrary, I have argued, they were going to die anyway, but they lost the possibility of immortality to which they came close. But, this being so, even on this new plane the limitations of humanity, of which death is a central symbol, become all the more clear: for the power of knowledge which has been gained implies a contact with the eternal and transcendent, but human limitations including death conspire to make that knowledge ineffective. Human life, wrongly pursued, can turn out to be like the existence of the beasts. 'Man cannot abide in his pomp, he is like the beasts that perish', is reiterated by Psalm 49.13, 21 [12, 20]. Adam, struggling as an agriculturalist with his refractory soil, would not have too much time to develop his intellectual faculties. In this more indirect sense we may find it possible to restate the traditional belief that through Adam we got into a situation whereby death came to have dominion over us.

4

Noah's Ark: Time, Chronology and the Fall

The reader may well ask what Noah and Noah's ark have got to do with a title like *The Garden of Eden and the Hope of Immortality*. Noah was never in the Garden of Eden, and most people have never thought of him in connection with immortality at all. But there are certain reasons why Noah and his story should come into our discussion. Here are some of them.

First, in Mesopotamian mythology, it is with the story of the flood and the hero who was saved in his boat that the symbol of the tree of life, or plant of life, appears. 'The Epic of Gilgamesh,' writes Stephanie Dalley, 'narrates a heroic quest for fame and immortality, pursued by a man ... of strength and weakness who loses a unique opportunity through a moment's carelessness.' These are close to the themes which we have already identified in the biblical story of the Garden of Eden. Utnapishtim knows of the plant and in the end of the story he is taken up into the world of the gods. Addressing Gilgamesh, he says:

> Let me reveal a closely guarded matter, Gilgamesh,
> and let me tell you the secret of the gods.
> There is a plant, whose root is like camel-thorn,
> Whose thorn, like a rose's, will spike [your hands].
> If you yourself can win that plant, you will find
> [rejuvenation (?)].[1]

There are many differences in the biblical story of Noah, but it is impossible to question that there is a relationship here.

A second such piece of evidence lies in the name of Noah himself. We have already seen, in the case of Eve, that a name given and explained in the Bible might reach back to much older stories or older stages of the same story, and this is the case with the name Noah also. The explanation of the name given in the Genesis text (זה ינחמנו: 'this one shall bring us relief' RSV) is transparently a folk-etymology and cannot fit the name Noah. G.R. Driver long ago suggested that the name Noah was cognate with the Ethiopic verb *noḥa*, 'be long (of time)', and corresponded to the Mesopotamian Ziusudra of another form of the flood legend.[2] It cannot be said that this is certain, and yet it is extremely fitting as an explanation, where no other that carries any conviction has been suggested.[3] It should be accepted, even if only tentatively. If I am right in this, in its earlier stages the story of Noah may have been yet another story of one who came near to eternal life or at least was in touch with the sources from which it might be obtained. This is probable in any case, just from the similarity of the Hebrew and the Babylonian flood stories, and this explanation of the name Noah only makes more intelligible what was intrinsically likely.

Moreover, taken in a wider perspective, the story of Noah is extremely relevant for all discussions of the beginnings of the world and the nature of humanity. Although on the surface of the text the world was created long before Noah, and Adam is our first ancestor, in a number of ways the world in which we live is a world that had its beginning with Noah and his times. To take the most obvious aspect, that idea of a sudden and catastrophic fall into sin and radical alienation from God, which has been characteristic of much of the Christian tradition and which was supposed to have its textual location in Adam and Eve and the Garden of Eden, actually has a much better textual basis in Genesis 6. It is here that the structure of the world is seriously violated by the angel marriages, in which the sons of God or heavenly beings mate with human women, producing mighty or monstrous beings who come somewhere between the two basic categories of purely divine and purely human, two categories which, on usual Hebrew assumptions, ought not in any way to be confused. It is here, and not in the story of the Garden of Eden, that we hear that the earth was filled with violence (Genesis 6.11).[4] It is here, and nowhere else in the Bible, that we find a statement coming close to the idea of 'total depravity': 'the Lord saw that the wickedness of man was great in the earth, and that the whole formation of the thoughts of his heart was only evil all the time' (Genesis 6.5). It was then, and appropriately, that God, disgusted with this state of affairs, regretted that he had made humanity at all and decided to wipe out the whole lot of them, which he did, keeping alive only one family, that of Noah. For Noah and the others, surviving a year on the ark among all these animals, and starting a new life on a virtually new world, all this was a highly unusual experience, amounting to not much less than a new start of life, or something like eternal life. It is not surprising, therefore, if later writers, like the writer of I Peter, took it that the story of the flood was closely suggestive of baptism on the one side and of resurrection on the other ('being put to death in the flesh and made alive in the spirit', I Peter 3.18–20).

'Not just for them but for us also', that same writer commented. The world in which we live is the world that began with Noah. We often comfort ourselves with the words of Genesis 1 to the effect that God saw that everything that he had made was good, and no doubt that was true, but does it still apply to the world as it is?[5] In the beginning God made a firmament or floor to keep out the chaotic waters, but later he opened it up to let them back in. After Noah's time, he said that he would not do this again; but who knows if he will not do something else of the same kind? This was a fear that already occurred to New Testament people, who pointed out (II Peter 3.5–10) that he might send destruction by fire next time, an even worse fate, and not an unrealistic one, as we well know today. The world of Genesis 1 was a vegetarian world, in which (it seems) lions and tigers lived on grass, and animal flesh was not to be eaten by mankind. One of the big changes in the post-deluge world was that permission was given for humans to eat meat: 'the fear and dread of you shall be upon every beast of the earth' (Genesis 9.2); and with this there goes one other big alteration in the constitution of the world, in that the shedding of human blood is to be avenged by the shedding of blood in requital, something that, apparently, had been unknown in the original constitution of the world. So the world as we know it goes back to Noah. There is no one here or anywhere who is not his descendant. From his family there came all the nations peopling all regions of the earth. Certain very basic rules of justice promulgated to him are supposed to be binding upon all.

Traces of this sort of character of the flood can be seen outside Israel. In Athens during the festival of the Anthesteria there was a day on which a special dish called *panspermia* was cooked, and the celebrants remembered that the few survivors of the mythological flood had eaten this dish as their first food.[6] Jan Bremmer writes: 'The Flood was associated with the social chaos of the Choes [Χόες or 'Wine-jugs'] on the mythological level, and the restoration of the social order on the Chutroi [Χύτροι or 'Pots'] is therefore appropriately connected with the survivors of the Flood. In Greece, as elsewhere, the Flood was often thought to have preceded the origin or foundation of important constituents of the cosmic and social order ...' Among these, interestingly, the Greeks included the mixing of wine: 'one of the inventors of mixed wine was Amphictyon, the son of Deucalion'[7] – the comparison with Noah and his drunkenness is not to be missed.

Putting it in another way, the story of Noah and the Flood functions as a sort of second creation. Some of the more poetic descriptions of creation have a closer resemblance to the recovery of the earth after the Flood than they have to the story of creation in Genesis 1. Thus Ps. 104:

> Thou didst set the earth on its foundations so that it should not be shaken;
> thou didst cover it with the deep as with a garment;
> the waters stood above the mountains.
> At thy rebuke they fled ...
> the mountains rose, the valleys sank down ...
> thou didst set a bound which they should not pass;
> so that they might not again cover the earth,

and similarly Psalm 74.12–17. In it God divided the sea, he crushed the monster Leviathan, he opened springs, but he also dried up everflowing rivers: I take this to mean that he made openings (by which floodwaters could run away beneath the earth) and dried up the streams of the cosmic river Ocean (wherever these still flowed over the earth). Thus he fixed the boundaries of the earth, the passage of seasons. This has more likeness to the Flood story, to the draining away of the waters and the drying up of the land, than it has to the traditional creation story.

It is, therefore, likely on all these grounds that the story of Noah and the Flood may give us good insights into the Bible's ideas of the basic constitution of the world and human nature, and into these we are now to look.

First of all we should go back to a point that was already mentioned, namely the role of the generation of the Flood as the uniquely serious locus of sin and evil in the beginnings of the world.

As most Christian tradition since St Paul has seen it, the disobedience of Eve and Adam, and of Adam in particular, was a totally catastrophic outbreak of evil, a drastic rebellion against God which involved all humans afterwards in sin and death. This, however, as we have seen, is far from a sound exegesis of the story of the Garden of Eden. In fact, it was rather with Genesis 6 that these catastrophic aspects came to be noted. The angel marriages, whereby the angelic 'sons of God' mated with human females and produced mighty giants, are little noticed by the modern reader, but for the ancient Hebrews these were of paramount importance, and it is no exaggeration to say that as an example of catastrophic evil they considerably excelled the doings of Adam and Eve. Correspondingly it is natural that God reacted with great violence, wiping out the entire human race except for eight persons, plus the total animal population (apart, presumably, from fish and aquatic animals), and overwhelming the entire world in water. Or, to put it in another way, if we are to talk of a 'Fall of Man' and indeed of a 'fallen world', we come much closer to it here in the Flood story than in that of the Garden of Eden. But, if it is a Fall in that sense, it works in a different way: for all the 'fallen' humans of this Fall were wiped out, while those who survived, being Noah and his family, were 'unfallen'; and since all of us are descended from that group of eight persons, none of us are affected by this Fall after all.

But we have to say more about the angel marriages. If the account given of the angel marriages in the canonical book of Genesis is brief and rather colourless, the same was not true of those who meditated on the passage and expanded it in the deutero-canonical literature, notably the Books of Enoch and of Jubilees. The rebel angels, and the giants whom their matings with women produced, were a shocking crew of evildoers and blasphemers. They were, indeed, technological innovators of a kind, such as we have already noticed to be true of the offspring of Cain, but their innovations were far from beneficial to mankind: among them there were three in particular which were reprehensible, for they originated (*a*) astrology, (*b*) armaments for the shedding of blood, and (*c*) cosmetics for the decoration of women.

> And Azazel taught men to make swords, and daggers, and shields and breastplates. And he showed them the things after these, and the art of making them: bracelets, and ornaments, and the art of

> making up the eyes and of beautifying the eyelids, and the most precious and choice stones, and all kinds of coloured dyes. And the world was changed. And there was great impiety and much fornication, and they went astray, and all their ways became corrupt. Amezarak taught all those who cast spells and cut roots, Armaros the release of spells, and Baraqiel astrologers, and Kokabel portents, and Tamiel taught astrology, and Asradel taught the path of the moon.[8]

Not unnaturally they had to be imprisoned in a dark cavern far beneath the earth to keep them out of the way. In Jesus' time they were still there and were the 'spirits in prison' to whom he was said to have preached (I Peter 3.19). The whole story is one of black rebellion and retribution, far more drastic and extreme than the disobedience of Eve and Adam. So, to conclude this point, some part of that load of concepts of sin and depravity which has been attached to the first human pair belongs more naturally to the generation of the flood and should be seen as belonging there.[9]

But, this being so, something in the region of life and perhaps eternal life belongs here, too. As I have already hinted, the value of life, the positivity or negativity of death and the concept of eternal life are connected with *time*. Now though many people of modern times, whether 'liberal' or 'conservative' in their attitudes to the Bible, have ceased to be aware of it, the subject of chronology and the reckoning of time is one of the major interests of the Old Testament.[10] Concerning this interest, however, there is no person, with the possible exception of Abraham, who plays a more important part than does Noah. The story of the Flood is carefully timed and dated. Noah was 500 years old when he became the father of Shem, Ham and Japheth. It was 100 years later when the Flood came upon the earth. It was in the 600th day of his life, in the second month, on the seventeenth day of the month, that the downpour of the heavenly waters began. It was exactly five months later, on the seventeenth day of the seventh month, that the ark came to rest on the mountains of Ararat. It was on the first day of the first month of his 601st year that the waters were dried up and Noah opened the covering of the ark and looked out. It was in the second month, on the seventeenth day of the month, exactly one year after the deluge began, that the earth was (this time, really?) dry and God told Noah to disembark. So the whole episode is very precisely timed, much more precisely than most other narratives of the Hebrew Bible. Time, then, and the measurement of time, is certainly an interest here.

We are more concerned here, however, with years than with months and days. Genesis provided an exact chronology from creation down to the Flood, then on to Abraham and the descent into Egypt, and from there on other books provided data that carried the reckoning of time down to the Exodus, to the building of Solomon's temple, and to the time of the kings of Israel and Judah. From the beginnings down to the Flood the numbers are given by the ages of the fathers in the genealogies of chs. 5 and 11. Thus Genesis 5.3–5 informs us that Adam was 130 years old when he had his son Seth, through whom his descent was to be counted (this genealogy ignored Cain and Abel). Similar figures are given for each generation, and by simple addition they give us a figure of 1656 AM (Anno Mundi) for the coming of the Flood: that is to say, there were 1656 years from creation to the start of the Flood. These figures are substantially different in the Greek text of the Septuagint and in the Samaritan Hebrew text, but their figures, though differing, belong within the same general category.

Moreover, the same passage tells us not only that Adam was 130 years old when Seth was born, but also that he lived 800 years thereafter and had other sons and daughters. This pattern is continued throughout the series, and most of the antediluvians, the people who lived before the Flood, lived something in the region of seven, eight or nine hundred years. This is, of course, a long time. As everyone knows, the longest life was that of Methuselah, who made it to 969 years; Jared had 962, and Adam came only a short distance behind with his 930. But what are these figures supposed to mean?

One obvious clue is that, as we move down through the centuries, the life-span of people becomes shorter. After the Flood, in Genesis 11, it drops rapidly, from four hundred odd to two hundred odd, after Abraham to below two hundred, and Moses himself lived a mere 120. In later times seventy was considered a proper span of life and eighty was unusually old.

What we may perhaps see in this material is a conception of a different type from what people have been accustomed to see in the Bible: a conception of a *gradual* decline from a noble and excellent earlier state, marked by a steady decrease in the ages to which persons lived. Naturally the biblical figures were not entirely the original product of the Hebrews, and they are both analogical to the (much higher) Mesopotamian figures and likely to have been derived in some way from them or from a common source. Equally naturally, they did not necessarily mean, within Hebrew culture, what they had meant in their distant past origin. Within Hebrew culture, the noticeable thing is that several of the figures come close to a thousand but none reaches it. Adam reaches 930, seventy years short of that significant point, and the shortfall of seventy is likely to be meaningful. As time goes past, people live for shorter spans. The great time was in the beginning of the world. It is conceivable that a life of a thousand years was thought to count as constituting virtual immortality. In prophetic speech, which on occasion hinted back at the speech of primeval times, one could talk of 'thousands' of vines worth a 'thousand shekels' (Isaiah 7.23), just as one could talk of the child who would die 'a hundred years old' (Isaiah 65.20). Adam, if this is right, came little short of the millennium of life which would have constituted virtual or practical immortality.

But in order to follow out this line we are not dependent on the above interpretation, which has taken the figure of a thousand as a symbol of near-immortality. Whether this is valid or not, we have a similar result by a route that is not subject to the same uncertainties. For what the genealogical material surely displays is a conception of slow, gradual movement, stage-by-stage decline rather than sudden, total catastrophe. And one of the reasons for this is the sentiment that the beginnings of the world were a marvellous time, when there were great people and wonderful things happened.

Nowhere is this better illustrated than in the personality of Adam himself. Starting from the traditional Christian conception of Adam and original sin, one might have expected Adam to be portrayed as a foul wretch and miserable sinner, but in fact there are many traditions that go in the opposite direction. Certainly Adam made a bad slip, he got us all into trouble, and yet on the other side he was a tremendous fellow, very positively valued. All sorts of legends grew up, which depicted him in this way. What about the size of his body, for instance (a subject about which the Bible actually tells us little)? He was a huge person, his body was so big that it filled the whole earth, a fact that was made clear from his name itself, for the four letters ADAM stood for east, west, north, and south: ἀνατολή, δύσις, ἄρκτος and μεσημβρία. After his disobedience he was reduced to a mere hundred cubits in height.[11] Ben Sira in his 'praise of famous men' seems (ch. 44) to have begun by omitting Adam, and the first person he listed was Enoch, but at the end of the list, after Zerubbabel and before he came on to the nearly contemporary Simon son of Onias, he goes back to the start, as follows (49.14–16):

> No one like Enoch has been created on earth,
> for he was taken up from the earth.
> And no man like Joseph has been born,
> and his bones are cared for.
> Shem and Seth were honoured among men,
> and Adam above every living being in the creation.

There was something glorious about Adam. Traditions in this sense persisted in Christendom, fostered no doubt by the biblical texts which pointed this way. Adam, after all, even if he had sinned badly, had (presumably) passed on the highly essential story of his own downfall. The genealogy of his offspring made it clear that he had survived through many generations of them and thus had every opportunity to repeat and amplify the story. Since he lived on till 930 AM, he was still there, and still teaching, in the time of Enoch, Methuselah and even Lamech. Noah was the first of the antediluvians who had not known Adam personally, but he had no doubt got the story in authentic form from those who had. All this was familiar in the theological and cultural tradition of Christendom not only through the Reformation but up to the eighteenth century and even beyond. To Luther it was clear that all these people were fine teachers and theologians; they formed a sort of primeval apostolic succession. The Wittenberg theologian Deutschmann declared not only that Adam was a theologian himself, but that he was an orthodox Lutheran, his teaching being in full agreement with the Augsburg Confession and the Formula of Concord.[12] Only in comparatively modern times did this highly realistic treatment of the ancient chronological tradition come to be forgotten, even by the most conservative Bible-believers.

Noah himself had unusual figures for his chronology. While most of the others in the genealogy had their first son at something around 130 or 160, Noah was 500 years old, half a millennium, when he had his set of three: Shem, Ham and Japheth. This 500 can hardly fail to be significant. Another hundred passed before the flood began. After the flood he lived another 350, and he died at 950, again coming close to the thousand.

Another feature should be added here. We are talking about a scheme of things that is, broadly speaking and to use an unusual term, 'protological' (the reverse of eschatological). The basic revelation of God is back in the beginning of things. One such view is already seen in the supremely authoritative place of the Pentateuch: the five-book document leads up to Moses' death, and by that time everything that is worth knowing has been made known. But, following the same line even farther, some later traditions pressed even farther back, beyond Moses, to Enoch, to Adam himself. The beginnings of the world were the time when things were good and right. There then followed a slow, a very slow, decline. God reacts on human deficiency by cutting down on their time of life, just as in the story of the tower of Babylon he reacted against human ambition by confusing human language. Thus a difficult phrase of Genesis 6.3, though obscure, certainly belongs here. 'My spirit will not remain in a human being for ever: because he is mortal flesh, he will live only for a hundred and twenty years,' says REB, in the verb following LXX οὐ μὴ καταμείνῃ, though there is some uncertainty about this meaning for the verb ידון. But what follows is clear: God is cutting down the life-span that can reasonably be expected for humans.

Where then does this element of our discussion lead us? It seems to lead us in two (or more?) different directions. On the one hand, our study of the traditions concerning Noah and the flood, and concerning the antediluvians and their times and ages, has confirmed our argument about the specific story of the Garden of Eden: the question of eternal life was an active one informing the early stories of Genesis and continued in activity through the reading of these stories. On the other hand, these same traditions have a very considerable rootage in legend and mythology, and at least in part are inherited from backgrounds outside Israel. They are certainly transformed within Israel, but they are not so transformed that they have lost the traces of their earlier life. Thus Noah in Genesis as it stands is no longer a seeker for the tree of life, which is not expressly mentioned in his story at all, nor is there any express word of his seeking immortality or eternal life. His story as it stands in Genesis is painted to look more like a highly unusual historical experience of escape from a cosmic catastrophe. Nevertheless, it seems hardly possible to explicate the story without recourse to the sort of unexpressed mythological hinterground which we have discussed.

What we have seen in the story of Noah reinforces one of the points that have been made, namely that the question of eternal life or immortality, as seen in these ancient texts, is very much a matter of *time*, time that can be measured in a chronological sense. Time is, if anything, more important than death. Questions of whether death is to be overcome, in the sense of people coming to life after death, are not prominent. Nor is there a serious concern with problems about the soul and the body.

The angel marriages are also an interesting example of another general aspect of stories about the origins of the world and humanity. Such stories come close to the mythological. This opinion has often been opposed by more conservative religious readers, but the angel marriages well show what is involved. Even the most conservative readers, enthusiasts as they are for the principle that one 'cannot pick and choose' within the Bible, are seldom heard to proclaim the importance of this incident. Since early times, as I have written in an earlier article, 'no major current of Christianity found much use for the angel marriages, over which men of biblical times seem to have laboured rather seriously, and they were allowed to gather dust in the lumber room of outdated and meaningless legend. As an explanatory device within theology, they had never worked very well; they did not give an account of anything within experience, and even to talk about them strained belief even more than most of the things that were believed.'[13] People did not have room, within their theology, for this sort of primeval negative incarnation in which semi-divine beings mated with human women, and therefore they ignored it or demythologized it by making it into an ordinary human and historical incident, as both Luther and Calvin did. For they took it as a report of how members of the true church (children of Seth and thus 'sons of God') unwisely intermarried with women from the false or Cainite church (children of Cain and thus 'daughters of men'), as ridiculous a piece of demythologization as ever was thought of. Calvin knew of the correct understanding, which was reported in patristic sources, but rejected it as 'gross ravings'.[14] In fact his theology had no tradition of any such negative incarnation and no place for it: since he personally did not believe the story, he denied its obvious meaning. In the Bible and the early post-biblical stages of Judaism, on the other hand, it was of enormous importance.

Part of its importance lay in its emphasis on *violence* as the fundamental sin. As I have emphasized, the Old Testament material seems, for the most part, to look on death in relation to *time and circumstance* rather than on death as a metaphysical generality. Some later comments, by contrast, seem to look on death itself, whatever the time or circumstance, as a terrible thing. Hebrews 2.14f. speaks of Jesus' purpose 'that through death he might destroy him who has the power of death, that is, the devil, and deliver all those who through fear of death were all their lives subject to bondage'. Paul's interest in the universality of death, inherited from Adam, seems to belong to the same world of ideas. This is an aspect of that Greco-Roman world, namely that one could think of death itself, death generally and universally, as feared, independent of the time and circumstance. Fear of death seems to be an almost metaphysical force. Here we can perhaps learn something of interest from these ancient legends of the time of the flood. For the evil that prompted the flood and brought it about was not death in itself, but *violence*, death in certain circumstances. Violence, in Hebrew *ḥamas* or חמס, was a characteristic term of those who concentrated their interest in this early period. It was with violence that the earth was filled, and this violence brought in the Flood as punishment and destruction.

This thought is interesting because it raises questions about the traditional Christian attitudes to death. A typical paradox of traditional Christianity has been its almost metaphysical insistence that death, all death, was unnatural, being introduced through sin, combined with its tacit moral acceptance that *war* was (after 'the Fall') virtually natural, being brought about by the perennial sinful state of humanity. Actually the reverse may well be the position of some major strata of the Bible, such as those we are looking at. Death, just death in itself, any death, is not at all unnatural. But violence, a particular accompaniment of certain very numerous cases of death, is indeed repugnant to God and contrary to his will. The metaphysical generalization of death, and the consequent failure to differentiate between one kind of death and another, has had very serious moral consequences in our tradition. Some at least of these prophetic passages in which God declares his opposition to death, his will to overcome it, belong to that same prophetic current which expresses a disgust with war and a longing for peace: Isaiah 2.4, the parallel in Micah 4.3, Isaiah 11.7, or still more 11.9.

Cullmann in his important booklet insists at considerable length on the *horror* of the death of Jesus. Unlike Socrates, who regarded death as a friend, whose death is a beautiful death, Jesus 'shares the natural [human] fear of death'.[15] 'Death for him is not something divine; it is something dreadful ... He is afraid of death ... Jesus knows that in itself, because death is the enemy of God, to die means to be utterly forsaken ...'[16] And so on. And these elements are by no means to be neglected or ignored. But in frankness it seems to me that this argumentation is neither honest, nor realistic, nor fully biblical.

First, it seems unrealistic to insist to this degree on the horror of the death of Jesus. That crucifixion is a terrible death is beyond doubt. But it can hardly be said realistically that for sheer horror the death of Jesus equals, or even comes close to, that of Jews in Nazi Europe or that of many thousands of soldiers and civilians, male and female, in the world wars, or that of political prisoners in many lands in modern times. And in saying this I am not expressing only my own judgment, for the Gospels themselves, though they certainly portray the factual horror of the scene, do not seem to stress this side to the extent that Cullmann's argumentation does. As they see the situation, Jesus knows what he is doing: he made a choice, he has a purpose, he is confident that his Father's will is behind all this. He knew that he was in the right. Not all who die in terrible circumstances have these elements of strength to support them.

And secondly, in any case, whatever we say of the death of Jesus, it is unthinkingly passive to suppose that what is true of his death is true of 'death' in the abstract, of all death. Of course it is not. It is, precisely, a death by violence, and in particular by enormous injustice: exactly the conditions under which the Old Testament *did* see death as something like an 'enmity to God'.

This, I believe, may be very important. Christian ideas about human immortality and future resurrection are in considerable measure ideas that are worked out to cope with the belief in Jesus *as saviour*, to cope with the need that is felt for *atonement*. Since it is understood that all need atonement, and since atonement is supposed to be through death, it is natural to connect the matter with the universal reality of human death. But this may be in part a premature universalization of something that in itself was more particular. I have in mind the length, the detail and the particularity of the passion narratives of the Gospels. That Jesus' death is the culminating point is obvious. But why such length and detail, when so many other stories about Jesus are tantalizingly brief? Could I suggest that the point is not the general, universal fact that he died, but the *way* in which he died – the victim of a complicated set of injustices, betrayals, insincere compromises, political deals and military brutalities? The way in which this happened was the point, and is transmitted by the narratives. It was death by violence, conspiracy and injustice. Something of this is lost when the matter is generalized into the universal of death and the universal of life.

I do not dispute that these ideas on which Cullmann based so much, the idea that death was the final opposition to God and so on, are to be found in the New Testament. But, disastrous to his argument and his whole outlook on these matters, exactly these ideas which he takes as central are likely to derive in some considerable measure from that very Greek philosophical tradition which he so abominates. As we have seen, the idea that sin brought death into the world is likely to be the product of the Hellenistic interpretation of the Old Testament, rather than the self-understanding of the Old Testament texts themselves; and still more, the idea that people 'through fear of death were all their lives subject to bondage' builds upon a metaphysical attitude to 'death' which is far distant from the Hebrew Bible. There is no Old Testament passage that expresses such a thought or comes near to it.

Since I have interpreted Noah's Flood, and the incident of the angel marriages which preceded it, as something like a 'Fall of Man', and indeed more a 'Fall of Man' than the narrative of Adam and Eve, I may conclude this chapter with some general reflections about the theological problems of such a 'Fall'.

In 1926, at the starting point of much that became regulative in twentieth century theology, the Swiss church periodical *Kirchenblatt für die Reformierte Schweiz* published a lively debate, in which the main protagonists were the Old Testament scholar Ludwig Köhler and the dogmatician of rising reputation Emil Brunner.[17] The subject of their argument was that same story of the Garden of Eden which we have been discussing. Recent theological discussion, Köhler said, had had a lot to say about original sin. In view of this, he felt it necessary to emphasize that this narrative could not be understood in dogmatic terms but was an aetiological myth. Its purpose was to explain a series of contemporary phenomena: why the snake has no legs, but creeps on its belly and eats dust; why there is enmity between humans and snakes; why women have pain in childbirth and have a desire for their husbands; why the soil is accursed; why humans wear clothes, and why they were driven out of Paradise. Only as a result of the Pauline 'Adam-Christ speculation' had all this become the basis for the churches' idea of 'original sin'. Köhler included several arguments that have been mentioned also by me, for instance the absence of any appeal back to the story of Adam and Eve within the Old Testament, and the absence of any interest in it on the part of Jesus. As a further argument, for good measure, he added that for Paul it was essential that the story should be taken as 'a historical fact', and that this was followed also by the Reformers. If one in the slightest degree questions or denies this historical linkage, he maintained, the entire Pauline doctrine of salvation is shattered and the Reformation doctrine of original sin is destroyed.

Brunner, not easily to be persuaded by this sort of talk, retorted that if Paul was no better than a novice in biblical interpretation, he personally would rather sit at the table among the novices, with Paul at their head.

In this rather unedifying discussion, there was something to be said on both sides. Aetiological explanations of the story of Adam and Eve were somewhat unsatisfying. They tended to insist that the story came into being in order to explain things which, for the ancient Hebrews, may well have been rather obvious facts, scarcely requiring any explanation in the first place, such as why snakes had no legs, why women have pain in childbirth, and so on. With some justification, aetiology of this kind seemed to people to be reductionistic: instead of ending up with some profound insight into the world and humanity, it produced some rather obvious point about some physical peculiarity. On the other hand, this was not really an objection against aetiological explanation as such. If someone objected to the categorization of the story as aetiological, one might have retorted that the traditional Christian understanding was aetiological too. For, according to it, the story existed in order to explain why people are the ghastly lot of evildoers that they obviously are. This, too, is an aetiology but on a higher level. It is an aetiology of sin and death. By contrast, the aetiology of the then Old Testament scholars, like Köhler, did not seem to promise any very profound theological insights. It offered an aetiology of rather obvious features of the human condition and environment, rather than an aetiology of profound and otherwise mysterious theological realities. For this reason it did not commend itself very widely. My own interpretation, however, if considered as an aetiology, does not suffer from this criticism. It is not an aetiology of universal human sin, nor one of death in the sense of an explanation of how death entered the world.

But it is none the less an aetiology of a profound theological reality, an aetiology of immortality, or more correctly of non-immortality. It tells why man comes close to God in respect of knowledge but remains mortal (as he always had been) in spite of that nearness. It is thus not an aetiology of minor and obvious features of life, but one that accounts for deep and mysterious realities. I accept, however, that some aetiological interpretations may seem theologically jejune, as the ones described above no doubt seemed to Brunner.

Brunner for his part, however, was also on very shaky ground. It was all very well to say that he was going to sit with St Paul. But he did not even try to resist the serious *exegetical* arguments put forward by Köhler, for there was no way in which he could have done so. Admittedly, Köhler's aetiological explanations had weaknesses and could be turned aside: but his arguments – which I also have put forward in this book – that the Hebrews never used the story of Adam and Eve in order to explain sin and evil, and that Jesus never used it in that sense either, were unanswerable. It was all right to scoff at the theological naivety of Old Testament scholars, mixing in some insulting imputations of Pelagianism by the way, but Brunner betrayed his own inability to provide any solid scriptural basis for his views from within the Old Testament text. In effect his position meant an abandonment of the evidence of the latter: Brunner implied, though he could not say it explicitly, that Paul and his understanding must be right, whether the Genesis text supported them or not. In fact, as I have shown, large elements in the text cannot be made to support Paul's use of the story without distortion of their meaning. And, as I have indicated, it is not hard to suggest an alternative and intermediate position: Paul was not interpreting the story in and for itself; he was really *interpreting Christ* through the use of images from this story. If the Old Testament text is to count as having some sort of authority in and for itself, then it must be free and able to utter a message of its own which may, at least in principle, be substantially different from the use which Paul made of certain selected and very limited elements within it, read through the perceptions and assumptions of a later and very different culture. It is useless to talk of the 'authority' of the Old Testament if in fact it is not to be allowed to say anything different from what Paul, or any other particular later interpreter, supposed it to be saying. What Brunner was in effect saying was that he was determined to have 'the Fall of Man' and 'original sin', whether the Old Testament text supported them or not. And in that he spoke for a great deal of modern traditionist theology.

As we noted, Köhler pointed out that for the Pauline and the traditional Reformational doctrine of the Fall and original sin, it was essential that the Fall of Adam and Eve should be a historical fact. Brunner's answer to this[18] was in essence a lecture on what he, Brunner, considered to constitute 'history'. 'History', according to him, began with and from the Fall ('*Die geschichtliche Welt – das was wir so kennen und nennen – beginnt diesseits des Urstandes als Folge des Falles*'). What Brunner unfolded was an interesting and quite sophisticated, but entirely modern, set of ideas. They formed no sort of answer to Köhler's argument, because they were not the ideas of Paul, or of the Reformation, or of the older theological tradition at all. Brunner had no strong sense of the historical.[19] The ideas of the Reformers were quite different from what he put forward. For them history began very definitely with creation, and Adam's life and Fall came within its framework. In this respect the 'liberal' Köhler knew the Reformers better than the neo-orthodox Brunner did. Certainly, therefore, as Köhler claimed, the older views of the Fall and original sin had been built upon the assumption of a historical Fall, and if this latter was no longer to be thought of as a historical fact, then these views could not fail to require modification.

The problems underlying these disagreements are probably still with us. Much theology, we may suspect, may be dimly aware of what Old Testament scholars think about Genesis 2–3, but finds it difficult to undertake an open reformulation of the traditional doctrines, especially in the aspect that I have taken as central, namely the mortality of Adam before the Fall and the consequent naturalness of death. For these are not just ideas about an ancient text, but have important repercussions on our understanding of the human condition now and at all times. The thought that *all* death, at all times and in all circumstances, is due to a primeval fault is difficult to take seriously, and all the more so when we perceive that Old Testament scripture by no means supports this idea. Similarly, the belief that God really, on the ground of a fault committed by two humans in the beginning of the world, ordained death as a destiny for all later humanity, throughout history, has truly staggering effects on the idea of God.

Now modern Old Testament scholarship has, I think, been agreed that the Pauline use of the story (we should call it a use rather than an interpretation) does not fit with the actualities of the text. In that sense, even if Brunner may have brushed aside Köhler's judgment in 1926, competent scholarship has continued to conclude that, at least on the negative side, Köhler was right. The story of Adam and Eve was not the story of the 'Fall of Man' in the traditional sense. At the end of a survey of mainstream modern scholarship, Sibley Towner pronounces it as 'univocal in all the literature of the main stream' that:

> There is no Fall in scripture, if by Fall one means the doctrine of the shattering of the divine image in humankind, the loss of immortality at an early moment in human history, and the inexorable transmission of the original sin through human genes ever after. There is no account of the origin of evil and no primeval encounter with Satan.[20]

But what then was the purport of the story? When we leave aside the rather puny aetiological understandings mentioned above, we find in many interpreters a very general, acceptable, but also rather vague expression. 'Seen as a whole,' writes Gene Tucker, 'the purpose of the narrative is to explain human existence, and particularly its present broken state.' The story is, in its broadest sense, 'an attempt to explain the fact that human beings experienced a tension between their own existence and some other reality or possibility, between what they are and what they are meant to be, or could have been'.[21] And this is quite right. 'Some other reality or possibility ... what they could have been': this is just what I have argued. But my interpretation has, I believe, made sharper and more precise just what this 'other reality or possibility' was. That other reality or possibility was focussed as immortality: immortality was the issue, and it formed a central starting point from which many other ideas grew. Indeed, one may say, Paul's own interpretation or misinterpretation, following the Wisdom of Solomon's same line of thought, can be seen to be a right recognition of this fact even if it is mistaken in its handling of the Genesis text. Immortality was the issue, and humanity ended up being (or remaining) mortal: Wisdom, followed by Paul, and later followed by the main theological traditions, rephrased this so as to say that the humans had been immortal but had lost this immortality. As I have put it, they never had it, but they had the chance of it, and lost that chance: this is not so far away.

As Towner says, having shown that there was no Fall in this story, it remains 'difficult to think of humanity without a Fall'.[22] Quite so. But the idea that humanity is 'fallen' does not depend entirely on the story of Genesis 3. If people believe in 'the Fall', in 'original sin' and the like, they do so not because of what they have read in Genesis, but on other grounds. These may be grounds of deduction from other accepted facts and principles. Paul himself reasoned in such a way: for example, 'we are convinced that one has died for all: therefore all have died' (II Corinthians 5.14). If one can reason in such a way, then one can certainly find confirmation for it, or exemplification of it, in Genesis. In the modern world, I think, the grounds are most of all grounds of experience: experience of failure, of the recrudescence of evil. 'Original sin' provides a formula under which this experience can be classified. Even if there were no story of Adam and Eve, one could swiftly come to the idea of a 'Fall' by such a procedure. The idea of a 'Fall' in Christianity rests on a much broader basis than the story of Adam and Eve. For most of Christianity, the story of the garden of Eden has been, not the ground on which the idea of a Fall was based, but the mode through which that idea was symbolized.

But why should it be difficult to 'think of humanity without a Fall'? Surely only because it has been supposed that humanity was originally perfect. But why should humanity have been originally 'perfect'? Genesis 1 said that everything God had created was 'good', but that comes short of total perfection and sinlessness. 'Perfection', once again, was read into Genesis 1 because of the Pauline typology of sin and death. In Genesis itself there is no such scheme. What God pronounced to be 'good' was the created world as it still is, the humans being mortal as they still are. Being 'good' means that God was entirely satisfied with what he had created. That world contained elements, as we have seen, like the waters, the darkness, which he had not created, but the whole as it emerged was 'good'. Humanity, as it was still much later, after the Flood, was still 'in the image of God', but it was hardly 'perfect'. The 'goodness' of creation and of humanity applied to the world that traditional theology regarded as 'fallen'.

The fact is, it seems, that humanity was never 'perfect', and therefore the idea of a 'Fall' is otiose. More likely, we have to think, they were imperfect from the start. This did not prevent God from regarding them, rightly, as a 'good' creation. Their imperfection makes it all the more natural that they disobeyed: everyone did. If we, today, are to imagine that we have a continuity with the humans of the beginnings of the world, we had better stop thinking of them as persons who possessed 'perfection'.

On the other hand, it might be argued, the loss of a potentiality is all that 'Fall' was ever intended to mean. We may leave aside the idea that Adam was immortal and became mortal; but, even so, if he had even the faintest chance of immortality and was firmly deprived of that, this is still an event which was highly determinative for human life. Perhaps the 'Fall' can be thus re-expressed. But if it is expressed in this way the accent and the emphasis of the idea become considerably different.

5

Immortality and Resurrection: Conflict or Complementarity?

One interesting feature of the modern discussion of our subject is the amount of passion and even anger that it can arouse. Cullmann begins his preface by saying that 'No other publication of mine has provoked such enthusiasm or such violent hostility'. A reader wrote to him saying that, when he was dying for lack of the Bread of Life, Cullmann had offered him nothing but stones and even serpents. Another reader took him to be a kind of monster who delighted in causing spiritual distress. 'Has M. Cullmann,' he wrote, 'a stone instead of a heart?' Stendahl got equally into trouble. 'I received more angry mail about it than about any other speech or article,' he says. No subject, it seems, is more perilous for discussion than immortality; danger money should certainly be paid to those willing to talk about it. Stendahl says that he can understand why the subject makes so many people angry, but he does not explain what that reason is. For his own part, he thinks, all he has done is to approach the subject 'on the basis of cool, descriptive biblical theology'. Cullmann, I am sure, thought the same thing.

Now something has to be said about this anger. The human race can perhaps, if we leave aside agnostics and scoffers, be divided into two great classes, the immortalists and the resurrectionists. Though neither of these scholars says it, it seems likely that the anger expressed against them came more from immortalists than from resurrectionists. It was the immortalists who were suffering in this debate. Both scholars, though in very different ways, wanted to cut down all ideas of the immortality of the individual soul, and to emphasize the much greater centrality of the resurrection. And if people were made angry by this, it was because they had received the impression, from long decades and centuries of religious teaching, that the immortality of the soul was central to Christian belief and morality. To be told that it was no part of true religion, but an inheritance from pagan philosophy and erroneous philosophy at that, seemed pretty hard. They reacted with anger.

Stendahl seeks to cope with this by suggesting that the immortalist position was flawed because it depended on an excessive interest in *individual* immortality. People, he thinks, are interested, rather self-centredly, in immortality for *themselves*, for 'little me', as he puts it, an expression he repeats rather too much. On the other hand, they want to have a secure knowledge of what happens after death, which is more than we ought to desire:

> Immortality and the concern for immortality appear much *too little*, too selfish, too preoccupied with self or even family, race and species ... And immortality is *too much* since it has a tendency to claim to know more than may be good for us.[1]

The real thing, he maintains, is something different:

> The question is not: What is going to happen to little me? Am I to survive with my identity or not? The question is rather whether God's justice will win out ... Resurrection answers the question of theodicy, that is, the question of how God can win, the question of a moral universe.[2]

Well, no doubt resurrection does answer these questions, but it does not follow that immortality is so *merely* a matter of selfish interest in one's own survival. For it may be answered that the same is true of resurrection in any case, in so far as resurrection applies to the eventual destiny of the individual. The argument that resurrection answers the problem of theodicy directs us, primarily, towards the resurrection of Jesus himself. In so far as belief in that resurrection is expanded towards an interest in *our own* mortal bodies and what will happen to them, that interest seems liable to criticism as a selfish interest of 'little me' to the same degree as is the case with the immortality of the soul. And so we have to move towards a conclusion by considering again some of the aspects in which immortality and resurrection may be interrelated.

It is, in fact, easy to state the cause of the anger directed towards scholars like Cullmann and Stendahl. In diminishing, or even seeking to abolish altogether, all interest in the immortality of the soul they were denying what many religious people regarded as one of the most precious heritages of their faith. Not surprisingly, these people reacted with dismay or fury. But, we may ask, why was it that modern theology turned so strongly against this particular element of belief?

This great shift in theological perception was, primarily, part of the strong reaction against theological liberalism, represented most fully by dialectical theology. One of the weaknesses of liberalism, as it came to be seen, was a vagueness about the resurrection. What sort of event was it? Take as a typical and powerful example Harnack's treatment of the subject in a few pages of his *What is Christianity?* 'If the resurrection meant nothing but that a deceased body of flesh and blood came to life again, we should make short work of this tradition.' The New Testament itself distinguishes between the Easter message, the stories of the empty grave and the appearances of Jesus, and the Easter faith. The stress lies on the latter. The Easter faith can and should be maintained even where the Easter message is absent or unknown. The Lord is a spirit, says Paul, and this carries with it the certainty of the resurrection. The formulations of the Easter message became more and more confident and complete, but 'the Easter *faith* is the conviction that the crucified one gained a victory over death; that God is just and powerful, that he who is the firstborn among many brethren still lives'. Paul based his faith not on stories like that of the empty grave but on internal revelation 'in me', coupled with a 'vision'. The empty tomb was not as important as the appearances of Christ, but these latter are no sound foundation of faith, for they are prone to historical doubts and depend on a 'miraculous appeal to our senses'. Therefore:

> Whatever may have happened at the grave and in the matter of the appearances, one thing is certain: *This grave was the birthplace of the indestructible belief that death is vanquished, that there is a life eternal*... The certainty of the resurrection and of a life eternal which is bound up with the grave in Joseph's garden has not perished, and on the conviction that *Jesus lives* we still base those hopes of citizenship in an Eternal City which make our earthly life worth living and tolerable. 'He delivered them who through fear of death were all their lifetime subject to bondage,' as the writer of the epistle to the Hebrews confesses. That is the point. And, although there be exceptions to its sway, wherever, despite all the

> weight of nature, there is a strong faith in the infinite value of the soul; wherever death has lost its terrors; wherever the sufferings of the present are measured against a future of glory, this feeling of life is bound up with the conviction that Jesus Christ has passed through death, that God has awakened him and raised him to life and glory.[3]

In spite of its eloquence, this statement seemed to have many defects. In particular, it seems to lay little emphasis on the *actuality* of the bodily resurrection, and to leave open the possibility that a *faith* in resurrection would be equally valid even if nothing had happened to the body of Jesus at all. The way in which Harnack linked his view with the thought of 'the infinite value of the soul' seemed to confirm this. And notably lacking was any explanation of how the resurrection of Jesus could be complemented and fulfilled through resurrection of Christian believers also, on the lines of Romans 8.11: 'If the Spirit of him who raised Jesus from the dead dwells in you, he who raised Christ Jesus from the dead will give life to your mortal bodies also through his spirit which dwells in you.' The creeds stated: 'I believe in the resurrection of the body',[4] and Harnack seemed not to give adequate value to this. Granted the points that Harnack makes, about the spiritual and visionary nature of resurrection belief, and about the insecurity of a faith that is founded on the narratives of the 'message' as such, something that gave more positive value to the *bodily* resurrection was felt to be needed. The bodily resurrection, and that meant the bodily resurrection *of Jesus*, was felt to be the central issue and point of conflict.

The turn away from the immortality of the soul and towards the (more or less sole) affirmation of bodily resurrection appeared to be a major counterstroke to the position exemplified by Harnack. Humanity existed only in and with a body, and this applied to Jesus as to others. Even he could not do anything or be anything unless in a bodily incarnate form. Admittedly there were difficulties in explaining exactly what had happened in the incidents of appearances reported in the Gospels, but whatever had happened, it had to happen in and with a body. Cullmann, similarly, a generation later, seems to be arguing primarily against the position of Bultmann, which, though coming out of dialectical theology and different in other respects from Harnack's, again appears to see the main reality of the resurrection in the existential 'decision' of faith, and opposes as improper 'objectification' any ultimate reliance on 'positive' data such as empty tombs, reliable witnesses and the like.

The appeal to Hebrew thought was an important part of the turn away from immortality of the soul. Appealing to the Old Testament as it did, it gave the impression of being a shift in a conservative direction: for, once again, it had often been objected against liberal theologies that they undervalued the Old Testament, while a classic Reformed theology like Calvin's had placed the Old Testament fully on the same level as the New. Basically, it was an argument that in theological lingo is called 'anthropological': the nature of humanity is such that there is no separable 'soul', and this principle, made clear in Genesis when the first man 'becomes' a living *nephesh* (Genesis 2.7) rather than 'having' a soul, is confirmed by aspects of the New Testament and is regulative for the whole. In biblical terms, therefore, it makes no sense to talk of the immortality of the soul and therefore, in respect of the resurrection of Christ, much heavier emphasis must be placed upon the theological values of its bodily nature.

In part this was an attempted return to older orthodoxies, but one would be misreading it if one saw it only in this way. Equally, it was an assimilation to modernity. On the popular level, people were no longer so interested in their immortal souls. 'I think it is true to say that an increasing number of men and women are less and less concerned about the immortality of the soul, especially their own,' Stendahl wrote in 1972.[5] Indeed so. By the 1930s or so, theology had turned away from the alliance with idealism which had been powerful in many quarters. In philosophy, many currents seemed to have lost interest in anything like a 'soul'. They could perhaps stretch as far as talking of a 'mind', though even that was not very certain, but a 'soul' seemed a quite unnecessary extra. Moreover, there was a strong liking for 'holistic' views of human nature: it was not a compound of a physical body with an immaterial soul, but an interlinked unity with interactions between physical and mental aspects. Thus modern psychology could speak of the *psyche*, using the same word as the Greek Bible, but apparently it did not mean the same entity as the 'soul' of traditional Christianity had been. Moreover again, the soul and its immortality suffered from the reaction against natural theology. One of the obvious functions of the older natural theology was precisely this, that it professed to demonstrate, on grounds of reason as distinct from scriptural and other revelational proofs, the existence of the soul and its immortality. Most modern theology turned against all such endeavours. One reason was that they seemed no longer to work. Another was that, if they worked, they produced the wrong result, working against Christian belief rather than for it: for if they proved that all humans had a soul, and an immortal one, then by that same proof everyone enjoyed this happy result whether they became Christians or not. And, of course, Greek philosophy had commonly been perceived as the main source of traditional natural theology. And it was from Greek thought that ideas of the soul and immortality had particularly been inherited. To get rid of these ideas was thus a natural accompaniment to the attack on all natural theology. Thus, to summarize this point, much in the turn against immortality of the soul was not a return to the fountain-head of biblical evidence but a climbing on the bandwagon of modern progress – the very thing that was at the same time being excoriated when it had been done in liberal theology.

Nevertheless, the turn against the soul and its immortality continued, and to this day continues, to be represented as a move back towards the Bible, towards the Reformation, towards the central verities of traditional faith. But exactly here lies the explanation of the anger and passion of which Cullmann and Stendahl wrote. The approach to the subject in modern theology greatly understated and obscured the extent to which traditional Christianity, Protestant as well as Catholic, had committed itself to that very Greek-based, immaterialist and immortalist, viewpoint which theologians were now so anxious to sweep under the carpet. Becoming convinced, along with the rest of the modern world, that body and personality belonged very much together, and that the immortality of the disembodied soul had rather little attraction for anyone, they 'found' the more welcome 'totality' view of the human person in the Hebrew outlook and went on from there to insist on interpreting the New Testament accordingly. And this could be quite justifiable, so long as it is not depicted as a mere restatement of traditional Christianity. On the contrary, traditional Christianity had invested far more heavily in the idealist, immaterialist, side of Greek philosophy than people now wanted to admit.

The Westminster Confession, for instance, a powerfully influential document of English-speaking Protestantism, took it as quite obvious that the human soul had 'an immortal subsistence'. 'Their souls (which neither die nor sleep), having an immortal subsistence, immediately [after death] return to God who gave them.'[6] This fact hardly needed to be proven;[7] to doubt it would probably have been a deadly heresy. And, because it was taken for certain, it at once affected the entire eschatological picture, since the soul had to live on without a body until the general resurrection, when the two had to be reunited again for the final judgement. Moreover, the immortality of the soul well symbolized and accompanied the conviction of the primacy and superiority of the immaterial over the material, which was generally understood to be one of the central teachings of Christianity. All this was not quickly to be forgotten.

In his earlier lectures in this series, to which I owe so much, Simon Tugwell has well described for us the problems of the different eschatologies, 'two-stage and three-stage', as he calls them, that might eventuate within the context of traditional Catholic theology. A 'two-stage' system he defines[8] as one where there is 'this life and the hereafter', a 'three-stage' system as one where there is 'this life, the hereafter, and, as it were, the thereafter'. Behind this stands the apparent ambiguity of the New Testament itself about eschatological expectations. On the one side there are indications that the righteous go straight into eternal communion with God, the key passage being the word to the thief on the cross: 'today you will be with me in paradise' (Luke 23.43); on the other side there is the clear insistence on a final judgment or 'Great Assize', to take place *after* the coming of the Son of Man, and, it was thought, after the general resurrection, clearly explained in Matthew 25. One way for the church to cope with this was to set them in temporal order: first there was one, then the other, over a wide temporal scale. In Catholic theology there was sense in this, in that there were ideas that some at least might utilize the intervening time in adding to their merits and improving their prospects for eternity. Since Tugwell's thoughts are very helpful in assisting us to organize our understanding, it seems good to provide a Protestant complement at this point by saying a little more about the position of the Westminster Confession, as a good example of older Protestant orthodoxy.

That the soul had 'an immortal subsistence' is not only stated by it but taken by it as obvious. The Confession is quite clear about what happens when someone dies. The body perishes but the soul, being of an immortal subsistence, lives on, and *immediately after death* is subjected to judgment for the deeds done in the bodily life, and consigned to heaven or hell as the case may be. Clearly souls were not so immaterial that they would not suffer pain (especially through heat!). Later, at the general resurrection, the soul is reinstated in the risen body, in order to take part in the great final judgment of all. But that judgment, it seems, though an enormous and impressive event, makes no difference, according to the Confession, to anyone's eternal status. Evildoers, whose souls have already been in hell since the moment of death, now go back there with their bodies as well, and conversely the same happens to the righteous.

In this context the importance of the immortality of the soul is obvious. Upon it depended the insistence upon the *immediate* judgment of the individual and the consequent denial of the Roman doctrine of purgatory, a denial which to Calvinists was a highly essential article of faith. After death absolutely nothing could be done that would affect anyone's eternal destiny. The soul went straight to heaven or hell and nothing would ever alter that. Reunion with a reconstituted body would be a sort of completion; the Confession is sure that they will be 'the self-same bodies, and none other, though with different qualities': those of the just will be raised 'to honour' and made 'conformable' to Christ's own glorious body, and those of the unjust will be raised 'to dishonour'. But, on the essential question of heaven or hell, bliss or damnation, there is no indication how this made any difference. Reunion with the body would, at the most, only confirm, and perhaps intensify, what had already been decided. Immortality might or might not be a Platonic idea, but no one worried about that, because it provided an essential plank in the Protestant platform.

And this was not all. The immortality of the soul provided a valuable defence against another possible error, namely the idea that the human being might be extinguished by death, and might thus escape some of the pains of eternal punishment. People might fancy this opinion, which would encourage them to have a good time in this world, knowing that they would be snuffed out at death and nothing more would happen. Not at all: the 'immortal subsistence' of the soul ensured that it would continue in existence for ever, and there would be no temporal limit to its pains of suffering. Thus, to sum up this point, the immortality of the soul was an essential part of much traditional Protestant religion, and retained this position down to quite recent times.

In this connection it is interesting to compare the Confession itself with an insightful, capable and sympathetic commentary from modern times such as is provided by my former colleague George S. Hendry in his book *The Westminster Confession for Today*. Hendry's fine work discusses the topics we have been considering, and in a balanced way considers the possible varying emphases on immortality of the soul and resurrection of the body.[9] But the reader of his work cannot fail to be puzzled by the apparent contrast between two different purposes: (*a*) to explain what the authors of the Confession themselves thought, and (*b*) to provide a modern commentary which would state, in modern form, the right theological view of these same topics in a mode that was sympathetic to the Confession and sought continuity with it but that was in fact considerably different from it. This difference, amounting at times to full contradiction, is very evident in the areas we have been discussing.

For example, Hendry, in the spirit of modern Reformed theology, emphasizes that 'we do not know' much, or anything, about life after death or its details. Christian hope is not founded on specific predictions of things to come. We have nothing but 'hope in Christ'. Quite so. As an expression of his own attitude to the problem, and of that of many, perhaps most, modern Christians, this may be entirely right. But it is not, it seems to me, what the authors of the Confession thought. They had no such modest idea. They knew very precisely what was going to happen, and said so. They might have agreed that it all amounted to 'hope in Christ', but, as they saw it, 'hope in Christ' provided in effect a very precise programme of future events. Not everything was known indeed: the Confession is wisely reticent about the millennium and such matters. But the essential blueprint is absolutely clear: the person is immediately judged after death; the soul continues to exist because of its 'immortal subsistence', and is consigned to its eternal habitation, in heaven or hell as the case may be. Later the soul is reunited with the body for the final judgment, but there is no indication of what difference this may make: all that happens is a reaffirmation of the judgment already passed at death, and the soul, now reunited with the body, goes back to the eternal abode in which it had already been since death. All this is very precisely known and defined, and no possibility of any alternative view is countenanced.

Dr Hendry gravely and wisely sums up the case as between the resurrection of the body and the immortality of the soul, as modern debate has analysed them.[10] But he does not make clear, as it seems to me, the essential point about the Confession itself: that for it there was no debate at all. The immortality of the human soul was just as certain and absolute an element in basic Christian belief as was the resurrection of the body, as absolute indeed as the resurrection of Christ himself. The reality of the immortality of the soul is assumed from the beginning and is an absolute necessity to the entire outlook of the Confession. No one had the slightest notion that the immortality of the soul might be considered, on ground of biblical testimony or of Christian theology, to be in some sort of contradiction to, or rivalry with, the resurrection of the body.

Indeed, if anything, it seems that the immortality of the soul was the more important and the more influential aspect. The (Stoic?) principle that the soul must control and subdue the demands and passions of the body was central to traditional Protestant ethics, and many New Testament passages were interpreted in this sense. That Jesus had risen bodily from the dead was of course true, and essential; but for the individual believer the relation of his or her soul to God was central. It was about the soul, primarily, that God cared; and it was through the soul that men or women knew and enjoyed God. And this was true for Calvin himself: the soul had primacy, the body was 'the primary source of human wickedness'.[11] That his attitude to the body was 'unbiblical' is well portrayed by Bouwsma. And, in respect of the resurrection of Jesus, he 'complained about the tendency of Christians to be more impressed by this carnal event than by the raising of the soul'.[12]

Nothing makes this more clear than the depiction by the Confession of the Great Assize or Last Judgment. As Hendry rightly says,[13] the vivid passage of Matthew 25 made it necessary that this be given full prominence: this is what would happen 'when the Son of Man comes in his glory'. But the solemn treatment of it by the Confession succeeds in reducing it to little more than a comic farce. For all humans have already been judged and sentenced immediately after death, and their souls are at once in bliss in heaven or in torment in hell. All that happens at the Last Judgment is that they are, through the General Resurrection, reunited with their bodies, and then back they go to heaven or to hell, as the case may be. Bodily form, it seems, does nothing to change any situation; at the most it serves to intensify the pleasures or pains of eternal bliss or punishment. And the Last Judgment, thus seen, has no drastic effect on anyone, other than what they had coming to them anyway.

It should, then, be obvious that in traditional Protestantism of this type there was no conflict or contradiction between resurrection of the body and immortality of the soul. If one was essential, so was the other. The bodily resurrection of Jesus was, naturally, essential for dogma and for the central soteriological doctrines. For the relation of the human person to God, however, the role of the soul, as the organism which was specially apt for contact with God and communion with him, and the primacy of soul over body, which was essential for innumerable points of ethics, and the immortality of the soul, which was essential for the *immediate* consignment of the person to eternal heaven or hell – all this was of entirely equal centrality. Indeed it was, if anything, more important.

It is of interest here to say a little more about the position of John Baillie, who was already mentioned in Chapter 1. As I said there, his book concentrated very strongly on the immortality of the soul, and was written in a very Platonic style: just what was, so soon afterwards, to come to be widely abominated. Though the resurrection was assumed, the primary interest was in the soul. This did not mean, however, as has been implied by so many critics of immortality, that the interest was selfishly personal and individual. On the contrary, as Baillie made clear, if there was a step beyond individual immortality and a greater prize to be gained, it was a *social* and *communitarian* perspective. This was, I think, his way of explaining what remained to be added through the eventual reunion of the soul with the body. Baillie insists, with italic for emphasis, that '*There can be no complete consummation for the individual until there is consummation also for society.*' This, he says, 'is the real significance of the conception of the "Last Day" '.[14] He then continues with a passage that deserves quotation at length:

> The orthodox teaching, both Roman and Protestant, is that until the last day the souls both of the blessed and of the damned remain *disembodied*, though already dwelling in what is to be their final place of abode; but that on the Last Day there will be a General Resurrection whereby the souls of both are reunited to their old bodies. This of course had to be taught, so long as it was also taught that in the life eternal we are to have the same bodies of flesh and blood and bones (though rendered incorruptible) that we now possess ... But the true significance of the teaching comes out only when we are told that the spirits of the blessed, though

now enjoying the heavenly bliss *substantially*, will not possess it *in the fulness of its accidental nature* until after the Last Day; for the deepest reason of this delay is not really that until then they will be disembodied but that until then their society will be incomplete.

He then goes on to quote, characteristically, Baron von Hügel's saying: 'it is not enough that individuals are immortal'. Immortality of the soul in the older tradition, then, did not necessarily involve petty individual self-interest.

I repeat, then: immortality of the soul had a deep personal and ecclesial grounding, and much of people's loyalty to religion was explicitly attached to this particular belief. It is not surprising, therefore, that scholars like Cullmann, who considered the resurrection of the body and the immortality of the soul to be antithetical, ran into considerable opposition. They were denying not only a tradition of Greek philosophy, but something that was built into the entire tradition of classical Protestantism. It was the modern, 'holistic', body-centred and resurrection-centred trend that was the innovation. This was the reason for the anger. Perhaps the present work, which suggests that the Bible was more interested in the soul and in immortality than modern trends have wished, will provoke equal or greater anger in the opposite direction.

What then of the issue as between immortality and resurrection, which Cullmann and many others after him made so central? He thought that these two concepts were mutually contradictory: anything you had of the one would damage or even destroy the other. But Simon Tugwell, among his many fine insights, has pointed out that the 'problematic' insisted on by Cullmann seems never to have existed in the New Testament or in early Christian literature: and to this we may well go on to add, in the texts of intertestamental Judaism as well. No one can be found to have denied resurrection on the grounds that they believed in the immortality of the soul. On the contrary, the immortality of the soul was invoked, when it was invoked at all, as a *support* to the belief in resurrection, or as a means of working out its implications which had been left uncertain, or – as I have suggested – as a stage in the movement of thought that actually worked to *produce* the novel idea of resurrection in the first place. For – to quote only one reason among several – if a body was brought to life, how was one to be sure that it was the same person, and not a *replica* of his or her body, unless a continuity on a non-bodily level existed? Conversely, it does not appear that the existence of Greek philosophy in itself formed an obstacle to the acceptance of resurrection, while conversely the acceptance of Jewish tradition did not provide a necessary avenue to its acceptance or a necessary reason for doubt about immortality.

One example that may occur to the reader is the reaction to Paul's speech on the Areopagus in Athens (Acts 17). It was when he came to mention the raising of 'a man' from the dead that some of his Greek audience began to 'mock'; and since the modern contrast between resurrection and immortality became fashionable it has been easy to think that this was the cause of their scorn: being Greeks, they must have believed in the immortality of the soul, and found talk of a resurrection absurd. There is nothing in the passage, however, to justify this idea. There is no indication that their mockery arose because they believed in the immortality of the soul. As I pointed out long ago,[15] in so far as any philosophers were supposed to be present in the background of that story, they were Stoics and Epicureans, neither of whom would have been firm believers in the immortality of the soul. In any case, though Paul talked with philosophers in the Agora, there is no report of their presence on the Areopagus. Stoics would have been interested, not so much in the totally *separate* immortality of the soul, as in the priority of the will over the passions and weaknesses of the body. The stolid local administrators of the Areopagus, if they laughed at this part of St Paul's argument, are more likely to have done so for a quite different reason: given his long and eloquent argument from a natural theology standpoint,[16] why would anyone believe that this was clinched and proved through the raising from the dead of a remote person of whom none of them, apparently, had heard, and of whom the speaker had, as far as our account of the incident tells us, given them no information at all? Paul's discourse may have amazed people for the reason that is there specified, namely that they were expecting a manifestation of crazy Oriental cults. The question of the immortality of the soul simply did not arise. In any case, in this respect these Greeks were quite like the Jews, most of whom likewise, in spite of their very different mental background, saw no reason why they should be convinced by accounts of this particular resurrection. At that time resurrection was still not a regular or compulsory belief for Jews, as the New Testament itself makes clear. If the Sadducees did not believe in resurrection, it was not because they believed in the immortality of the soul; and even for those who believed that resurrection was to come it by no means followed that they should therefore believe in the accounts of any particular alleged resurrection.

The whole issue of Greek philosophy, which has been set in such violent opposition to biblical thought, is not so much to be overcome by reconciling the two, as by observing that the matter is simply not very important. As I have argued elsewhere, the contrast of Hebrew (or biblical) and Greek thought was no more than a part of twentieth-century theology's attempt to create its own image in the culture of the ancient world. There were indeed reasons why biblical thought stayed aloof, for the most part, from the Platonic tradition in these matters, but the reasons were different reasons from the ones that have been so over-emphasized in modern discussion. For one thing, the Socratic/Platonic argument for immortality of the soul depended heavily on the *eternity* of the existence of the soul, before as well as after, indeed more before than after, since knowledge depended on the *previous* acquaintance with the forms, which in this life were *remembered*. This was distant from the Hebraic background, though one cannot say that it was not combinable with its tradition in due course.[17] For another, as Tugwell well indicates, the most frequent collision between the Jewish/Christian and the Platonic traditions came over another matter altogether, namely the possible transmigration of the soul and its reincarnation in a different person or body altogether, an idea which did not at all suit either orthodox Judaism or Christianity. Yet in some strands of Judaism even this view was found viable and attractive: Josephus seems to ascribe it to the Pharisees.[18] There were therefore substantial reasons why the Jewish/Christian tradition could not invest too heavily in the Platonic tradition of argument, even if it accepted a good deal of the *conclusions* of that argument.

I began these lectures by giving what may have been a modest boost to immortality, making it the central feature and theme of the story of the Garden of Eden, and connecting it also with the significance of other figures of the early story such as Enoch and Noah. It seems to me undeniable that this view has strong support in the detailed matter of the Genesis text. But this approach has several ramifications. First of all, if it is right, it not only places an interest in immortality of a kind right back in the beginning of our Bible; it also gives it a strategic position in which it is linked with many developments that follow later but have proved difficult to keep within a convincing historical account.

Secondly, however, it links this material and these ideas with the environing world of ancient religion and mythology out of which Israel and its traditions emerged. Thirdly, it alters the focus of the entire subject because it places before us a third conception: not the immortality of the *soul* (as apart from the body), and not the resurrection of the dead (after death, as contrasted with bodily life before death), but the continuance of life, a sharing in the kind of 'eternal life' which is the life of God, and the avoidance of the inevitability of death which is taken to be normal for humanity.

The first difficulty that such a conception will cause is clearly its mythological aspect. Both traditional resurrection and traditional immortality have a sort of physical realism about them. If a body comes alive again, we can get going with living once more. If it does not, then there may be a soul that keeps going independently. Either of these, or both, are for modern man, even modern religious man, more rational and easier to follow than the idea of just staying alive for ever, without death. Thus even the most voracious modern 'Bible believer', who vehemently insists on a literal physical resurrection of Jesus, keeps fairly quiet about the idea that Enoch is still up there, in his original body, just as he was six thousand (or six million) years ago. 'Eternal life' of this sort will be quick to receive extensive demythologization, most of all from those who in other regards most oppose any such interpretative programme. And in fact some such demythologization very possibly took place within ancient Israel, and this forms part of the route by which we come eventually to the idea of 'eternal life' as something marked by its quality rather than its temporal duration – the very aspect which Stendahl, and many others, have found to form the most positive aspect of biblical thought in the whole area.

But that 'quality of life' approach to eternal life was not the only path which biblical thinking followed. It led also to resurrection. Painting it on the broadest possible canvas, we might say: humanity lost that glimmer of a chance of immortality, and so mortality remained its boundary. Apart from extreme exceptions like Enoch, a future possibility lay only through death and not around it. Add to this the burden of human sin and guilt – for, even if Adam was not the culprit, plenty of it accrued over the centuries of human existence – and the need for atonement. We thus have in perhaps surprising degree a sort of imaginative link running back through the Old Testament and forward to the resurrection of Christ.

But do people want to have such a linkage? Relations here are paradoxical. Undoubtedly exegetical opinions are directed by apologetic priorities. Many conservative Christians will be consoled, momentarily at least, by the thought that a line leading towards eternal life, and perhaps even resurrection, is well established within the Old Testament, going back to Adam himself. Did not Jesus say that the Son of Man must be killed and after three days rise again, leaving a distinct impression that there was scriptural evidence for this? Surely the resurrection is a genuine fulfilment of all this line of expectation?

Unfortunately, this is a double-edged sort of argument. Undoubtedly the strong and insistent emphasis on the resurrection, as opposed to the immortality of the soul, within modern theology was stimulated by apologetic pressures. One no longer sought to work by rational arguments that would, from without the precincts of faith, establish that the human soul was immortal by nature. Resurrection was an event, a historical act within history, not an argument concocted on the basis of premisses and dialectics. This was the ultimate ground for Cullmann's impassioned argumentation: resurrection belonged to *Heilsgeschichte*, history of salvation; immortality of the soul was to be rejected with scorn because it was an appeal to eternal, immutable, timeless values. It was, in the modern world, easier to believe in the absolute miracle of the resurrection than to go through the slow process of considering the destiny of the soul and how and when it might be united with the body.

This was one side: but there is another. Resurrection seems at first sight to benefit from earlier biblical support: there was a line that led up to it, it was a fulfilment. But, paradoxically, the more such support it has, the more uncertain it becomes. For the more expectation there was that a great religious leader should come alive again after death, the more that same expectation goes to explain the claims that it had been fulfilled.[19] Resurrection, peculiarly, is at its strongest when it is an absolute, completely isolated, miraculous event. The more it is accompanied by assorted revivifications of persons in Old Testament and New, some of them to die again after their rising, plus the numerous saints who arose and appeared after Easter, to say nothing of the heavenly translations of Enoch and Elijah and perhaps others, the more doubtful it all sounds. Of course, such things do not *prove* anything definite about the resurrection, but they make a difference to the way in which it is likely to be regarded.

The effect of the modern turn against immortality of the soul has been to exercise a damaging pressure upon much understanding of the New Testament. Consider a statement like this by a respected and careful Catholic exegete:[20]

> For him [Paul], as throughout the Bible, the soul, created by God together with the body, is mortal as is the body. Actually it dies through sin. If God restores it to life through the forgiveness of redemption, it is not by setting free in it a life that it possessed naturally, but by re-creating that life which it had entirely lost.

How extremely positive and certain an assertion, in an area where considerable nuance and qualification might be desirable! Where is the evidence that for Paul the soul 'is mortal as is the body'? It seems to me most unlikely that he thought any such thing. I cannot find any statement in his letters that he thought the soul to have died along with the body. Perhaps, one may suspect, the statement is based on the remarks in I Corinthians 15.42–54 about different kinds of bodies, the σῶμα ψυχικόν and the σῶμα πνευματικόν. The former is 'sown', the latter is raised. But this does not necessarily entail that the soul 'dies' with the body. The scheme might be that the soul lives on and is transformed into the 'spirit' which actuates the transformation of the body. Other possibilities could be thought of. Unless we have definite statements that the soul dies with the body, the reverse possesses the overwhelming cultural probability: the soul lived on except where a clear negative is provided. 'Flesh and blood', says Paul in v. 50, do not inherit the kingdom of God: these are exactly the entities that do *not* include the soul. And where is the evidence that this is the case 'throughout the Bible'? The reality is the opposite: it is the conviction that there is a uniform view about this, existing 'throughout the Bible', that has caused Paul to be credited with an opinion which may well be the opposite of what he believed. For I have already quoted the words of Jesus himself, Matthew 10.28, advising one not to fear those who can kill the body but cannot kill the soul, and to fear rather the one who can destroy both soul and body in hell.[21] These words may imply that the soul is mortal in the sense that it *can* be killed (though not by ordinary human action), but they certainly do not mean that it is mortal in the sense that it simply dies when the body dies. Only someone of very special power can kill the soul: Cullmann says this is God himself, though to me it would seem obvious that it is the devil. Anyway, the death of the soul is something that does not easily and naturally happen: and that brings us a little closer to the immortality of the soul. If the writer of Matthew thought this, why could not Paul have thought it?

That resurrection is absolutely central to the New Testament goes without saying. But this does not in itself mean that immortality, or the immortality of the soul, is thereby denied, or marginalized. Immortality, it is sometimes said, is scarcely mentioned in the New Testament: there are only two occurrences of the noun ἀθανασία, people point out,[22] one of them about God's own immortality, which is beyond question, the other in 1 Corinthians 15.53 concerning how mortal nature must put on immortality. This last, Stendahl says, is perhaps not Paul's own thinking, but borrowed from other people whom he is quoting.[23] Maybe, but to me it looks like a principle that Paul takes as regulative – not something derived from Jesus' resurrection, but something that is independently known.

In any case the argument from numbers, even if quickly qualified, creates a quite wrong impression. Stendahl rightly says, indeed, that it is not a matter just of the number of the words; quite so. But, if one even mentions the number of occurrences at all, one has to go on and say that the figure of two is an understatement, even for the New Testament alone. For alongside ἀθανασία we have to include ἀφθαρσία and ἄφθαρτος, which increase the figure considerably: about six and eight cases respectively, with some question over variant readings. These look, at a rather literal glance, like 'incorruptible' and 'incorruptibility', and one might think that therefore they attach to flesh rather than to immaterial realities. But a quick review shows that this is not so. In most cases their actual meaning is in the field of 'immortal, immortality'. Thus Wisdom 12.1 says that, 'Thy ἄφθαρτον πνεῦμα is in all things' – no one supposed that a divine spirit was corruptible, and the point is that it is immortal. Similarly with κήρυγμα in Mark 16, the shorter ending, which may or not be original but certainly belongs to the biblical world. Romans 2.7 is likely to mean general 'immortality' rather than incorruptibility of the flesh, certainly so also Ephesians 6.24, certainly also some others. Immortality was certainly a significant concept in the New Testament.

Secondly, we have to include the highly significant figures of the books now considered (by traditional Protestantism) apocryphal. First, for anything involving the use of words we have to consider the apocryphal books: canonical or not, they are valid evidence of usage, and, as I have emphasized, at least some of these books are of great importance for the thought of writers like Paul. As soon as we look into the Wisdom of Solomon, or into III and IV Maccabees, we find plenty of evidence. The noun ἀθανασία occurs five times in Wisdom and twice in IV Maccabees, the adjective ἀθάνατος once in Wisdom and thrice in IV Maccabees. Similarly, we have ἀφθαρσία thrice in Wisdom and twice in IV Maccabees, and ἄφθαρτος twice in Wisdom.

The Bible, then, has much interest in immortality, as we have seen. Immortality of *the soul* is not the only kind that matters. Immortality of the person, by-passing death, is an important part of the heritage.

It is not surprising, therefore, that translations like KJV and RSV use 'immortality' as the rendering at a place like II Timothy 1.10: Christ 'abolished death and brought life and immortality to light'. Similarly, the various places that talk of the corruptible putting on incorruptibility are related to a place like I Corinthians 15.53 which tells that 'this mortal [nature] must put on immortality'.

And along with this, in spite of all the appeal that there has been to the Hebraic totality concept and the like, there is plenty of distinction between soul and body, body, soul and spirit, and the like. The 'anthropology' of one like Paul is very complicated, and I would not dare to attempt an account of it. But I can say one thing about it: to me, as one who has been a Hebrew scholar most of his life, it seems very alien from the Old Testament world. Moreover, nowhere is this more obvious than in his use of the terminology of the body and of resurrection.

For who will undertake to explain to us how σῶμα ψυχικόν and σῶμα πνευματικόν are characteristic concepts based within the usage of the Hebrew Bible? 'Body' (*sōma*) itself, the very central term of emphasis, hardly exists as a relevant term of the Old Testament: it is notorious that classical Hebrew has no such word, except marginally. Only later Hebrew, the usage of the Hellenistic period, develops one. נבלה is a corpse and cannot be used for a living body. The nearest characteristic term is בשר 'flesh'. It is interesting that some formulations of the creeds used this term: 'I believe in the resurrection of the flesh'. Commenting on this, Cullmann[24] says: 'Paul could not say that. Flesh and blood cannot inherit the

Kingdom. Paul believes in the resurrection of the *body*, not of the *flesh.*' But this argument, if taken seriously, only proves Paul's distance from the Hebrew Bible: the creeds that used 'flesh' were closer to it than he was. A glance at Septuagintal usage reinforces the point: in the Greek Old Testament, σῶμα was massively a dominant term of the apocryphal books. In the major areas of the Hebrew canon, if we leave aside places where it renders words for 'corpse', as in I Kings 13, it is at most sporadic: in Leviticus a group, mainly in ch. 15 and all concerned with ritual washings; otherwise only a handful. Daniel has some examples, its Aramaic having a normal term 'body'. The book in which σῶμα is most common is Job, with about a dozen cases, mostly having no clear Hebrew correspondent: that is, they come from the rather free composition style of the Job translator. Turn to the 'apocryphal' books, and we find: Wisdom five cases, Ben Sira eleven, Maccabees taken together about thirty! With his preference for σῶμα Paul was setting himself clearly in the context of Hellenistic Jewish usage.

It looks to me as if Paul had no idea that he was supposed to be working with Hebraic totality concepts or any other such thing. Obviously his sayings have indeed to be read with the Hebrew background in mind, but hardly in such a way that the Hebrew background (of earlier Old Testament times) can override what the texts are obviously uttering. I have already mentioned several possibilities: that Jewish thought had altered as against the ideas of early Old Testament times; or that quite a lot of the New Testament's view of humanity came (*horribile dictu*) out of the Hellenistic background anyway; or that Greek thought apart from the particular Platonic tradition had never been so diametrically opposed to Hebrew thought anyway. What is not acceptable is the attempt, in the name of a dubiously-constructed picture of Hebrew thought, to override rather obvious meanings of the Greek of the New Testament. The New Testament certainly says little directly and specifically about the immortality of the soul; but it has a reasonable degree of mention of immortality, and it certainly has an awareness that things of the body and things of the soul could take different directions. If the older theological traditions put two and two together, it was not so very surprising.

We have seen that the Old Testament and the following period of Jewish thought left various themes open: immortality of the soul, resurrection of some few, general resurrection, no resurrection at all, 'eternal life' earned and enjoyed here and now, and other eschatological schemes. The particular concentration on resurrection, characteristic of Christianity, appears late and seems to be associated with martyrdom.[25] It is not surprising that this is so when the peculiar message of Christianity is concentrated on one individual who suffered a death similar to martyrdom, who was followed by other martyrs, and who was believed to have been raised again to life by God. Such a concentration could well have left unsaid a large number of other conceptions which, just because they were cultural assumptions, were left unsaid. Moreover, as has been said above, it is a noticeable aspect of the ideas surrounding the martyrs that these ideas, more than any others, combined the thought of bodily resurrection with the thought of immortality of the soul, for reasons that were briefly sketched above.[26]

It was, perhaps, the practicalities of suffering and death that made this plain. Dead people remain as corpses. If there is to be a resurrection, some temporal scheme must govern what happens. But did the martyrs simply go out of existence until, at some distant date, their bodies were recreated? Hardly, and there was a strong tradition that those who passed away passed at once into the divine presence.

Another point may be added: although the resurrection of Jesus is, for Christians, the one central and dominant miraculous act of God, for the intellectual problems of our discussion it actually makes things somewhat easier. This is no doubt one reason why the resurrection became so much a centre of emphasis in modern times. The resurrection of Jesus is intellectually easier to cope with, for reasons of temporality: first of all, it is, for us all, in the past. It is over and complete, long ago. Secondly, the interval between death and resurrection was extremely short, less than forty-eight hours. No one had time to be wondering what had happened to him, or how the destiny of his soul was related to his body. This all the more so because, if we follow the Gospel narratives, no one gave a thought to the possibility of his resurrection, or even of some kind of more limited survival, until after the resurrection had happened. So it may be a challenge to faith but intellectually it is comparatively easy to categorize in respect of the questions we have been discussing. The problems become more severe as soon as we begin thinking of other people. The resurrection of Jesus is supposed to carry over into things that will happen to other people. But these other people are dead, or will die, and the general resurrection, we fear, lies chronologically a long distance away. Chronological time then comes back into our question once again.

Chronological time came into it in another way also. Part of the emphasis on the resurrection came from the idea that the time of the end was already upon us, and the general resurrection belonged to that time. For believers other than Jesus himself, therefore, there was no great problem, since in a quite short time the end of the world would come. 'There are some of those who stand here who will not taste of death until they see the kingdom of God come in power' (Mark 9.1; cf. Matthew 16.28; Luke 9.27). Such a point of view discourages worrying, either about individual resurrection or, even more, about the immortality of the soul. This makes it less surprising that the New Testament has little about such immortality. Passage of time, however, makes a difference. Later documents betray a worry arising from the failure of the Parousia to materialize. Others talk more about 'eternal life' as quality of life to be enjoyed now, but, as I have said, this in the end is likely to develop into thoughts of immortality. Once the church moved out from the New Testament period, the interest in immortality of the soul – even in a very Platonic sense, but with the proviso that the immortality is not intrinsic in the nature of the soul itself, but depends on the will of God – rapidly escalates.[27]

In traditional theology, Genesis 3 contained not only the story of the Fall: though the modern church consciousness, at least in Protestantism, has largely forgotten it, it contained also the *Protevangelium*, the First Promise, expressed in the words to the snake:

> I will put enmity between you and the woman,
> and between your seed and her seed;
> he shall bruise your head,
> and you shall bruise his heel (Genesis 3.15).

Sadly, according to modern scholars, there is no Protevangelium here, no promise of a future struggle with evil, no promise of final salvation. Calvin himself damped down such hopes.[28] Quite so. Nevertheless, according to the interpretation which I have followed, there may still be a Protevangelium in the passage, even if it is not in this particular saying addressed to the snake. The real First Promise in the passage lay elsewhere: it lay in the mention of the tree of life. The tree of life is made inaccessible to the humans; but it remains in existence. At some more distant time it remains possible that the way to the tree of life will once again be made available. From time to time, through the biblical texts, intimations of such immortality are to be felt.

> For all of you [says Ezra in IV Ezra 8.52–54]
> paradise lies open;
> the tree of life is planted
> the age to come is made ready ...[29]

And at the very end of the New Testament, in Revelation 22.2, exactly so the apocalyptist sees the tree of life, 'whose leaves are for the healing of the nations' – and note *nations*, rather than individuals and rather than 'the world' universally. Nations, we remember, are the units par excellence of *war*.

Immortality, then, was on the biblical agenda from the very beginning, with Adam and Eve. In the Garden of Eden there was the tree of life. The human pair might just have got to that tree, but they did not, because God stopped them; no one was to enter the Garden, and the cherubim with flaming sword stood there to guard the gate. Humanity was not fit to come near the tree. Nevertheless the tree remained there in the garden. Later one came to redeem the defect of humanity. Immortality was brought to light.

Bibliography

'Afterlife', in *Encyclopaedia Judaica* (Jerusalem: Keter 1971) II, 336–69 [author not specified: signed as 'ED.', which according to I, 16 means 'an internal editor'; one may guess, G. Scholem or Cecil Roth]

Aldwinckle, R.F., *Death in the Secular City* (Grand Rapids: Eerdmans 1974)

Allen, D.C., *The Legend of Noah: Renaissance Rationalism in Art, Science and Letters* (Urbana: University of Illinois Press 1963)

Badham, P., *Christian Beliefs about Life after Death* (London: SPCK and New York: Barnes and Noble 1978)

Bailey, Lloyd R., *Noah: the Person and the Story in History and Tradition* (Columbia: University of South Carolina 1989)

Baillie, John, *And the Life Everlasting* (London: Oxford University Press and New York: Scribners 1933)

Barr, J., *The Semantics of Biblical Language* (London: Oxford University Press 1961, reissued London: SCM Press 1983)

—, *Biblical Words for Time* (London: SCM Press 1969)

—, 'The Symbolism of Names in the Old Testament', *Bulletin of the John Rylands Library* 52, 1969–70, 11–30

—, *Old and New In Interpretation* (London: SCM Press 1966, 1982)

—, 'Philo of Byblos and his "Phoenician History"', *Bulletin of the John Rylands Library* 57, 1974–5, 17–68

—, 'The Authority of Scripture: The Book of Genesis and the Origin of Evil in Jewish and Christian Tradition', in *Christian Authority* (Henry Chadwick Festschrift, Oxford: Clarendon Press 1988), 59–75

—, 'Why the World was created in 4004 BC: Archbishop Ussher and Biblical Chronology', *Bulletin of the John Rylands Library* 67, 1984–85, 575–608

—, *Biblical Chronology: Science or Legend?* (Ethel M. Wood Lecture: London: Athlone Press 1987)

—, *Holy Scripture: Canon, Authority, Criticism* (Oxford: Clarendon Press 1983)

—, 'Luther and Biblical Chronology', *Bulletin of the John Rylands Library* 72, 1990, 51–67

—, ' "Thou art the Cherub" (Ezekiel 28.14)', forthcoming in a Festschrift soon to be published

—, *Biblical Faith and Natural Theology* (Oxford: Clarendon Press 1993)

Barrett, C.K., 'Immortality and Resurrection', *London Quarterly and Holborn Review* 190, 1965, 91–102

Barth, C, *Die Errettung vom Tode in den individuellen Klage- und Dankliedern des Alten Testaments* (Zollikon: Evangelischer Verlag 1947)

Barth, Karl, *Church Dogmatics*, Vol. III/2 (Edinburgh: T. & T. Clark 1960)

Benoît, P., and Murphy, R., *Immortality and Resurrection* (New York: Herder and Herder 1970)

Berger, Klaus, *Die Weisheitsschrift aus der Kairoer Geniza* (Tübingen: Francke 1989)

Birkeland, H., 'The Belief in the Resurrection of the Dead in the Old Testament', *Studia Theologica* 3, 1949, 60–78

Birkner, H.-J., 'Natürliche Theologie und Offenbarungstheologie: ein theologiegeschichtlicher Überblick', *Neue Zeitschrift für Systematische Theologie* 3, 1961, 279–95

Bloom, Harold (with David Rosenberg), *The Book of J* (New York: Weidenfeld 1990)

Bonhoeffer, Dietrich, *Creation and Fall* (London: SCM Press 1959)

Bouwsma, W.J., *John Calvin* (New York: Oxford University Press 1988)

Boyarin and Siegel, 'Resurrection, Rabbinic Period', *Encyclopedia Judaica* 14, 1971, 98–101

Braine, D., *The Reality of Time and the Existence of God* (Oxford University Press 1988)

Brandon, S.G.F., 'The Origin of Death in some Ancient Near Eastern Religions', *Religious Studies* 1, 1965, 217–28

Bratsiotis, N.P., '*Nepheš-Psychè*, ein Beitrag zur Erforschung der Sprache und der Theologie der Septuaginta', *VTS* 15, 1966, 58–89

Bremmer, J., *The Early Greek Concept of the Soul* (Princeton 1983)
Bridges, J.T., *Human Destiny and Resurrection in Pannenberg and Rahner* (New York: Lang 1987)
Burkert, Walter, *Homo Necans* (University of California Press 1983)
—, *Greek Religion* (Oxford: Blackwell 1985)
—, *Ancient Mystery Cults* (Cambridge, Mass.: Harvard University Press 1987)
Cadbury, Henry J., 'Intimations of Immortality in the Thought of Jesus', in Stendahl, (ed.), *Immortality and Resurrection*, 115–49; also in *HTR* 53, 1960, 1–26, and in *The Miracles and the Resurrection* (SPCK Theological Collections 3, London 1964), 79–94
Carnley, P., *The Structure of Resurrection Faith* (Oxford: Clarendon Press 1987)
Cavallin, H.C.C., *Life after Death. Paul's Argument for the Resurrection of the Dead in I Cor 15* (Coniectanea Biblica, New Testament series 7:1, Lund: Gleerup 1974)
Childs, B.S., 'Adam', in *Interpreter's Dictionary of the Bible* (Nashville and New York: Abingdon Press 1962), Vol. 1, 42–4
—, 'Death and Dying in Old Testament Theology', in Marks and Good (ed.), *Love and Death in the Ancient Near East*, 89–92
Collins, J.J., *Between Athens and Jerusalem: Jewish Identity in the Hellenistic Diaspora* (New York: Crossroad Publishing Co 1986)
Cullmann, O., *Immortality of the Soul or Resurrection of the Dead? The Witness of the New Testament* (London: Epworth Press 1958) – earlier in *Harvard Divinity School Bulletin* 21, 1955–6, 5–36, and later with an added Afterword, in K. Stendahl (ed.), *Immortality and Resurrection*; also in T. Penelhum (ed.), *Immortality*
Dahl, M.E., *The Resurrection of the Body* (London: SCM Press 1962)
Daiches, David, *God and the Poets* (Oxford: Clarendon Press 1984)
Dalley, Stephanie, *Myths from Mesopotamia* (Oxford University Press 1989)
Delcor, M., *Mito y Tradición en la Literatura Apocaliptica* (Madrid: Cristiandad 1977)
Dihle, Albrecht, 'Totenglaube und Seelenvorstellung im 7. Jahrhundert vor Christus', in Stuiber Gedenkschrift, *Jenseitsvorstellungen in Antike und Christentum* (Jahrbuch für Antike und Christentum, Ergänzungsband 9; Münster: Aschendorff 1982), 9–20

Doeve, J.W., review of Cullmann's *Immortalité de l'âme ou resurrection des morts?*, *Novum Testamentum* 2, 1958, 157–61

Edwards, Jonathan, *Original Sin* (Works, Vol. 3, ed. Clyde A. Holbrook, Yale 1970)

Eichrodt, W., *Theology of the Old Testament* II (London: SCM Press and Philadelphia: Westminster Press 1967), ch. 24, 496–529, 'The Indestructibility of the Individual's Relationship with God (Immortality)'

E.E. Ellis, 'Soma in 1st Corinthians', *Interpretation*, April 1990

Evans, C.F., *Resurrection and the New Testament* (London: SCM Press 1970)

Frost, S.B., 'The Memorial of the Childless Man: A Study in Hebrew Thought on Immortality', *Interpretation* 26, 1972, 437–50

Fuss, Werner, *Die sogenannte Paradieserzählung* (Gütersloh: Mohn 1968)

Gardner, Anne, 'Gen 2: 4b–3: A Mythological Paradigm of Sexual Equality or of the Religious History of Pre-exilic Israel', *SJT* 43, 1990, 1–18

Garland, Robert, *The Greek Way of Death* (Ithaca: Cornell University Press 1985)

Gerleman, G., 'חיה ḥjh **leben**', *THAT* 1, 549–57

Gestrich, Christof, *Neuzeitliches Denken und die Spaltung der dialektischen Theologie* (Tübigen: Mohr 1977)

Gnilka, J., 'Contemporary Exegetical Understanding of "the Resurrection of the Body" ', *Concilium* 1970/6, 129–41

Goldin, Judah, 'The Death of Moses: An Exercise in Midrashic Transposition', in Marks and Good, *Love and Death in the Ancient Near East*, 219–25

Grayston, Kenneth, *Dying We Live* (New York: Oxford University Press and London: Darton, Longman and Todd 1990)

Grelot, P., *De la Mort à la vie éternelle* (Paris: Cerf 1971)

Hebblethwaite, Brian, response to N. Lash in *Heythrop Journal* 20, 1979, 57–64

Heinzmann, R., *Die Unsterblichkeit der Seele und die Auferstehung des Leibes* (Münster: Aschendorff 1965)

Hendel, R.S., 'The Flame of the Whirling Sword', *JBL* 104, 1985, 671–4

Hendel, R.S., 'Of Demigods and the Deluge: Toward an Interpretation of Genesis 6: 1–4', *JBL* 106, 1987, 13–26

Hengel, M., 'Ist der Osterglaube noch zu retten?', *Theologische*

Quartalschrift 153, 1973, 252–69
Hick, John, *Death and Eternal Life* (London: Collins 1976)
Hillers, Delbert, 'Dust: some Aspects of Old Testament Imagery', in Marks and Good, *Love and Death in the Ancient Near East*, 105–10
Horst, P. W. van der, 'Pseudo-Phocylides and the New Testament', *Zeitschrift für die neutestamentliche Wissenschaft* 69, 1978, 187–202
Hügel, F. von, *Eternal Life* (Edinburgh: T.& T. Clark 1912)
Hughes, J., *Secrets of the Times: Myth and History in Biblical Chronology* (Sheffield: JSOT supplement series 66, 1990)
Humbert, Paul, *Études sur le récit du paradis et de la chute dans la Genèse* (Neuchâtel 1940)
Illman, Karl-Johan, *Old Testament Formulas about Death* (Åbo: Åbo Akademi 1979)
Jaeger, Werner, 'The Greek Ideas of Immortality', *HTR* 52, 1959, 135–47; also in: Stendahl (ed.), 97–114, and in Jaeger's *Humanistische Reden und Vorträge* (Berlin: de Gruyter 1960), 287–99
Johnson, A. R., *The Vitality of the Individual in the Thought of Ancient Israel* (Cardiff: University of Wales Press 1964)
Jónsson, Gunnlaugur, *The Image of God: Genesis 1:26–28 in a Century of Old Testament Research* (Coniectanea Biblica, Old Testament series 26, Lund: Almquist & Wiksell 1988)
Jüngel, Eberhard, *Tod* (Stuttgart: Kreuz Verlag 1971); ET *Death. The Riddle and the Mystery* (Edinburgh: St Andrew Press 1975)
Kaiser, Otto, 'Tod, Auferstehung und Unsterblichkeit im Alten Testament und im frühen Judentum – in religionsgeschichtlichem Zusammenhang bedacht', in O. Kaiser and E. Lohse, *Tod und Leben* (Kohlhammer Taschenbücher 1001; Stuttgart: Kohlhammer 1977), 7–80 and 143–57
—, 'Die Zukunft der Toten nach den Zeugnissen der alttestamentlich-frühjüdischen Religion', in his *Der Mensch unter dem Schicksal* (Berlin: de Gruyter 1985), 182–95
—, 'Der Tod des Sokrates', in his *Der Mensch unter dem Schicksal* (Berlin: de Gruyter 1985), 196–205
Kasher, A., and Biderman, S., 'Why was Baruch de Spinoza Excommunicated?', in D.S. Katz and J.I. Israel (eds.), *Sceptics, Millenarians and Jews* (Leiden: Brill 1990), 98–141

Kellermann, Ulrich, 'Überwindung des Todesgeschicks in der alttestamentlichen Frömmigkeit vor und neben dem Auferstehungsglauben', *ZThK* 73, 1976, 259–82

Knibb, Michael A., 'Life and Death in the Old Testament', in Ronald E. Clements (ed.), *The World of Ancient Israel* (Cambridge University Press 1989), 395–415

Krieg, Matthias, *Todesbilder im Alten Testament* (AThANT 73, Zurich: Theologischer Verlag 1988)

Lambert, W.G., and Millard, A.R., *Atra-Ȟasis: the Babylonian Story of the Flood* (Oxford: Clarendon Press 1969)

N. de Lange, *Origen and the Jews* (Cambridge: Oriental Publications 25, 1976)

Lash, N.L.A., 'Eternal Life: Life "after" Death', in *Theology on Dover Beach* (London: Darton, Longman and Todd 1978 and New York: Paulist Press 1979), 164–82

Laurin, R. B., 'The Question of Immortality in the Qumran Hodayot', JSS 3, 1958, 344–55

Levenson, J.D., *Creation and the Persistence of Evil* (San Francisco: Harper and Row 1988)

Lewis, H., *The Self and Immortality* (London: Macmillan and New York: Seabury Press 1973)

Lieberman, S., 'Some Aspects of After Life in Early Rabbinic Literature', in S. Lieberman (ed.), *Harry Austryn Wolfson Jubilee Volume* (3 vols., Jerusalem: American Academy for Jewish Research 1965), II, 495–532

Lifschitz, B., 'Inscriptions grecques de Césarée en Palestine (Caesarea Palestinae)', *Revue Biblique* 68, 1961, 115–26

Lifschitz, B., 'La vie d'au-delà dans les conceptions juives. Inscriptions grecques de Beth Shearim', *Revue Biblique* 68, 1961, 401–11

Lys, D., 'The Israelite Soul according to the LXX', *VT* 16, 1966, 181–228

Macquarrie, John, *Christian Hope* (Oxford: Mowbray 1978)

Manson, T.W. 'The Bible and Personal Immortality', *Congregational Quarterly* 32, 1954, 7–16

Marcheselli-Casale, C., *Risorgeremo, ma Come? Risurrezione dei corpi, degli spiriti o dell'uomo?* (Supplementi alla Rivista Biblica 18, Bologna: Dehoniano 1988)

Marks, J.H., and Good, R.M., *Love and Death in the Ancient Near*

East. Essays in Honor of Marvin H. Pope (Guilford, Conn.: Four Quarters 1987)
Martin-Achard, R., *De la mort à la résurrection, d'après l'Ancien Testament* (Paris and Neuchâtel: Delachaux & Niestlé 1956)
—, *From Death to Life* (Edinburgh: Oliver and Boyd 1959)
—, 'Trois remarques sur la résurrection des morts dans l'Ancien Testament', in his *Permanence de l'Ancien Testament* (Geneva, Lausanne, Neuchâtel: Cahiers de la Revue de Théologie et de Philosophie 1984), 170–84, and in *Cazelles FS*, AOAT 212, 1981, 301–17
—, 'L'espérance des croyants d'Israël face à la mort selon Esaïe 65, 16c–25, et selon Daniel 12, 1–4', in his *Permanence de l'Ancien Testament* (Geneva, Lausanne, Neuchâtel: Cahiers de la Revue de Théologie et de Philosophie 1984), 285–97
Masson, C., 'Immortalité de l'Âme ou Resurrection des Morts?', *RThPh* 8, 1958, 250–67
Matthews, Melvyn, 'Who Told You that You were Naked?', *Theology*, May/June 1990, 212–20
May, H.G., 'The King in the Garden of Eden: A Study of Ezekiel 28: 12–19', in B.W. Anderson and W. Harrelson (eds.), *Israel's Prophetic Heritage: Essays in Honor of James Muilenburg* (New York: Harper and Row and London: SCM Press 1962), 166–76.
Mayer, R., 'Der Auferstehungsglaube in der iranischen Religion', *Kairos* 7, 1965, 194–207
McEvenue, S., *Interpreting the Pentateuch* (Collegeville, Minnesota: Liturgical Press 1990)
Meyer, R., *Hellenistisches in der rabbinischen Anthropologie. Rabbinische Vorstellungen vom Werden des Menschen*, BWANT 76 (Stuttgart 1937)
Meyers, Carol L., 'Gender Roles and Genesis 3 : 16 Revisited', in *The Word of the Lord shall go forth* (Freedman FS; ASOR/Eisenbraun 1983)
—, *Discovering Eve* (New York: Oxford University Press 1988)
Moltmann, Jürgen, *The Crucified God* (London: SCM Press and New York: Harper & Row 1974)
Nickelsburg, G.W., *Resurrection, Immortality, and Eternal Life in Intertestamental Judaism* (Cambridge, Mass.: Harvard Theological Studies 26, 1972)
—, review of Stemberger, *CBQ* 35, 1973, 555f.
Nikolainen, A.T., *Der Auferstehungsglauben in der Bibel und ihrer*

Umwelt (Helsinki: Druckerei der finnischen Literaturgesellschaft 1944)

North, C.R., *The Second Isaiah* (Oxford: Clarendon Press 1964)

Otzen, Benedikt, 'Old Testament Wisdom Literature and Dualistic Thinking in Late Judaism', *VTS* 28, 1975, 146–57

Otzen, B., Gottlieb, H., and Jeppesen, K., *Myths in the Old Testament* (London: SCM Press 1980)

Pagels, Elaine, *Adam, Eve and the Serpent* (New York: Vintage Books 1988)

Parker, Robert, *Miasma* (Oxford: Clarendon Press 1983)

Pedersen, Johannes, *Israel* (2 vols., London: Oxford University Press 1926)

Penelhum, Terence (ed.), *Immortality* (Belmont, California: Wadsworth 1973)

—, *Survival and Disembodied Existence* (London: Routledge and Kegan Paul 1970)

Perrett, R.W., *Death and Immortality* (Dordrecht: Nijhoff 1987)

Phillips, D.Z., *Death and Immortality* (London: Macmillan 1970)

Ploeg, J. van der, 'L'immortalité de l'homme d'après les textes de la Mer Morte', *VT* 2, 1952, 171–5

Porter, F.C., 'The Pre-Existence of the Soul in the Book of Wisdom and in the Rabbinical Writings', in R.F. Harper, F. Brown and G.F. Moore (eds.), *Old Testament and Semitic Studies in Memory of William Rainey Harper* (Chicago: Chicago University Press 1906), 1, 207–69

Preuss, H.D., *Eschatologie im Alten Testament* (Wege der Forschung 480, Darmstadt: Wissenschaftliche Buchgesellschaft, 1978)

Quick, Oliver C., *The Christian Sacraments* (London: Nisbet and New York: Harper 1927)

Rahner, K., *On the Theology of Death* (New York: Herder 1961)

Reese, J.M., *Hellenistic Influence on the Book of Wisdom and its Consequences* (Analecta Biblica 41, Rome: Biblical Institute Press 1970)

Reventlow, Henning Graf, *Problems of Old Testament Theology in the Twentieth Century* (London: SCM Press and Philadelphia: Fortress Press 1985)

Richardson, N.J., 'Early Greek Views about Life after Death', in Easterling, P.E., and Muir, J.V., *Greek Religion and Society* (Cambridge University Press 1985), 50–66

Rosenthal, Franz, 'Reflections on Love in Paradise', in Marks and Good, *Love and Death in the Ancient Near East*, 247–54

Rost, L., 'Alttestamentliche Wurzeln der ersten Auferstehung', in W. Schmauck (ed.), *In Memoriam Ernst Lohmeyer* (Stuttgart: Evangelisches Verlagswerk 1951), 67–72

Rüger, H.P., *Die Weisheitsschrift aus der Kairoer Geniza* (Tübingen: Mohr 1991)

Saggs, H.W.F., *The Encounter with the Divine in Mesopotamia and Israel* (London: Athlone Press 1978)

Sasson, Jack M., 'A Genealogical "Convention" in Biblical Chronography?', *ZATW* 90, 1978, 171–85

Sawyer, J.F.A., 'Hebrew Words for the Resurrection of the Dead', *VT* 23, 1973, 218–34

Schunack, G., *Das hermeneutische Problem des Todes* (Tübingen: Mohr 1967)

Schwankl, Otto, *Die Sadduzäerfrage (Mk. 12, 18–27 parr)* (Bonner Biblische Beiträge, 66: Frankfurt: Athenäum 1987)

Schwartlander, J., *Der Mensch und seine Tod* (Göttingen 1976)

Simpson, J. Y., *Man and the Attainment of Immortality* (New York: Doran 1922)

Sparks, H.F.D., *The Apocryphal Old Testament* (Oxford: Clarendon Press 1984)

Spronk, K., *Beatific Afterlife in Ancient Israel and in the Ancient Near East* (Alter Orient und Altes Testament 219, Neukirchen: Neukirchener Verlag 1986)

Stemberger, G., *Der Leib der Auferstehung. Studien zur Anthropologie und Eschatologie des palästinischen Judentums im neutestamentlichen Zeitalter* (Rome: Biblical Institute Press 1972)

Stendahl, K. (ed.), *Immortality and Resurrection* (New York: Macmillan, 1958): includes introduction by Stendahl and essays by Cullmann, Wolfson, Jaeger and Cadbury

—, 'Immortality is Too Much and Too Little', in Stendahl, K., *Meanings. The Bible as Document and as Guide* (Philadelphia: Fortress 1984), 193–202

Sutherland, Stewart R., 'Immortality and Resurrection', *Religious Studies* 3, 1967, 377–89

—, 'What Happens after Death?', *Scottish Journal of Theology* 22, 1969, 404–18

—, *God, Jesus and Belief* (Oxford: Blackwell 1984)

Sykes, S.W., 'Life after Death, The Christian Doctrine of Heaven', in R.W.A. McKinney, *Creation, Christ and Culture* (Edinburgh: T. & T. Clark 1976), 250–71

Towner, W. Sibley, 'Interpretations and Reinterpretations of the Fall', in Francis A. Eigo, *Modern Biblical Scholarship: Its Impact on Theology and Proclamation* (Villanova, Pa.: Villanova University Press 1984), 53–85

Trible, Phyllis, *God and the Rhetoric of Sexuality* (Philadelphia: Fortress Press 1978 and London: SCM Press 1991)

Tromp, N.J., *Primitive Conceptions of Death and the Nether World in the Old Testament* (Rome: Pontifical Biblical Institute 1969)

Tucker, Gene M., 'The Creation and the Fall: A Reconsideration', *Lexington Theological Quarterly* 13, 1978, 113–24

Tugwell, S., *Human Immortality and the Redemption of Death* (London: Darton, Longman and Todd 1990)

Turner, James Grantham, *One Flesh. Paradisal Marriage and Sexual Relations in the Age of Milton* (Oxford: Clarendon Press 1987)

Urbach, E.E. *The Sages: Their Concepts and Beliefs* (Jerusalem: Magnes Press 1975)

Vawter, B., 'Intimations of Immortality and the OT', *JBL* 91, 1972, 158–71

Vischer, W., *Das Christuszeugnis des Alten Testaments*, I (originally published 1934), ET *The Witness of the Old Testament to Christ* (London: Lutterworth Press 1949; only this volume was ever published in German or English)

Wächter, L., *Der Tod im Alten Testament* (Stuttgart: Calwer Verlag 1967)

Ward, Eileen de, 'Mourning Customs in 1, 2 Samuel', *Journal of Jewish Studies* 23, 1972, 1–27 and 145–166

Wasserstein, Abraham, 'Die Hellenisierung des Frühjudentums. Die Rabbinen und die griechische Philosophie', in W. Schluchter (ed.), *Max Webers Sicht des antiken Christentums* (Frankfurt am Main: Suhrkamp 1985), 285–316

Westermann, C., *Genesis* 1–11 (Minneapolis: Augsburg Publishing House and London: SPCK 1984)

—, 'נֶפֶשׁ *næfæš* Seele', THAT II, 71–96

Wickham, L.R., 'The Sons of God and the Daughters of Men: Genesis vi 2 in Early Christian Exegesis', OTS 19, 1974, 135–47

Wijngaards J., 'Death and Resurrection in Covenantal Context (Hos 6.2)', VT 17, 1967, 226–39

Wiles, Maurice F., *The Remaking of Christian Doctrine* (London: SCM Press 1974)

Williams, C.J.F., 'Knowing Good and Evil', *Philosophy* 66, 1991, 235–40

Williams, N.P., *The Ideas of the Fall and of Original Sin* (London: Longmans 1927)

Wilson, R.R., 'The Death of the King of Tyre: the Editorial History of Ezekiel 28', in Marks and Good, *Love and Death in the Ancient Near East*, 211–18

Wisdom, John, 'Eternal Life', in *Talk of God* (Royal Institute of Philosophy Lectures, II, 1967–68, London: Macmillan 1969), 239–50

Wohlgschaft, H., *Hoffnung angesichts des Todes* (Munich: Schoningh 1977)

Wolff, Hans Walter, *Anthropology of the Old Testament* (London: SCM Press 1974)

Wolfson, H.A., 'Immortality and Resurrection in the Philosophy of the Church Fathers', in his *Religious Philosophy: A Group of Essays* (Harvard University Press 1961); cf. *Harvard Divinity School Bulletin* 22, 1956–7, 5–40

—, 'Notes on Patristic Philosophy', HTR 57, 1964, 119–31

Zimmerli, W., *Man and his Hope in the Old Testament* (London: SCM Press 1971)

Notes

1 *Adam and Eve, and the Chance of Immortality*

1 K. Stendahl, 'Immortality is Too Much and Too Little', in K. Stendahl, *Meanings. The Bible as Document and as Guide* (Philadelphia: Fortress Press 1984), pp. 193–202; quotation from p. 196. See also his Introduction to the volume of four Ingersoll Lectures at Harvard, *Immortality and Resurrection* (New York: Macmillan 1965). Stendahl, though agreeing with Cullmann in wishing to diminish or eliminate the immortality of the soul, was critical of many of the individual types of argument that were used by Cullmann and others who argued in the same way.
2 J. Moltmann, *The Crucified God* (London: SCM Press 1974), p. 170; cited by N. Lash, *Theology on Dover Beach*, pp. 171f. For a discussion of Moltmann's views, see S.W. Sykes, 'Life after Death, the Christian Doctrine of Heaven', in R.W.A. McKinney, *Creation, Christ and Culture* (Edinburgh: T. & T. Clark 1976), pp. 250–71.
3 W. Vischer, *Das Christuszeugnis des Alten Testaments*, 1, 266 (originally published 1934), ET *The Witness of the Old Testament to Christ* (London: Lutterworth Press 1949), 1, 217.
4 Ironically, exactly the same evaluation has been passed on Christianity: 'The world of the Homeric Greek gods owes its shining splendour to its remoteness from death; to educated Greeks Christianity was to appear as a religion of the grave' – so Walter Burkert, *Greek Religion* (Oxford: Blackwell 1985), p. 203.
5 See the Appendix 'The Resurrection of the Body', in Maurice F. Wiles, The *Remaking of Christian Doctrine* (Cambridge University Press 1974), pp.125–46.
6 Oliver C. Quick, *The Christian Sacraments*, p. 94; quoted by Wiles, *Remaking* (n.5), p. 127.
7 For an earlier and briefer approach to the same subject see my 'The Authority of Scripture. The Book of Genesis and the Origin of Evil in Jewish and Christian Tradition', in G.R. Evans (ed.), *Christian*

Authority (Henry Chadwick Festschrift, Oxford: Clarendon Press 1988), pp. 59–75.

8 The argument I present here is not at all a new one, but many have evaded its cogency by simply ignoring it. N.P. Williams, *The Ideas of the Fall and of Original Sin* (London: Longmans 1927), pp.115ff., tries to overcome it by arguing that the epistolary style leaves unmentioned things that were universally believed; but even he has to resile from the full implications of this argument, and rests content with the view that at least 'no other theory of the origin of evil was in possession of the Gentile-Christian field' at the time of Paul's writing.

9 In the ancient world, generally speaking, a god might be killed (by another god), though this seldom happened; but a god would not grow old and die 'naturally', i.e. of old age.

10 This is generally accepted. 'Must be Jacob, not Abraham', says C.R. North, for instance, comparing Hosea 12.3 (*The Second Isaiah*, Oxford: Clarendon Press 1964, p. 130). He does not even consider Adam as a possibility. A more serious candidate would be Aaron, as was kindly suggested to me by Dr J. Gerald Janzen. The context mentions sacrifices (Isaiah 43.23f.), and the 'princes of the sanctuary' are mentioned immediately afterwards (v. 28). Among the figures of early times in Israel, Aaron was outstanding as a sinner, having taken the initiative in the notorious incident of the Golden Calf.

11 W. von Soden, *Mitteilungen der Deutschen Orient-Gesellschaft* 96 (1965), 48; quoted by H.W.F. Saggs, *The Encounter with the Divine in Mesopotamia and Israel* (London: Athlone Press 1978), p. 117. On the whole subject see Saggs's ch. 4, pp. 93–124, 'The Divine in Relation to Good and Evil'.

12 Harold Bloom, *The Book of J* (New York: Weidenfeld 1990), p. 175.

13 S. McEvenue, *Interpreting the Pentateuch* (Collegeville, Minnesota: Liturgical Press 1990), p. 65.

14 Text follows that of R.H. Charles as revised by J.P.M. Sweet and published in H.F.D. Sparks, *The Apocryphal Old Testament* (Oxford: Clarendon Press 1984), p. 614. The term '250 *times*' may be understood as '250 weeks of years', following Sweet's note. 'Times' for years or longer periods goes back to Daniel.

15 Sifre Deuteronomy 357, p. 428. See the charming essay by Judah Goldin, 'The Death of Moses: an Exercise in Midrashic Transposition', in J.H. Marks and R.M. Good, *Love and Death in the Ancient Near East. Essays in Honor of Marvin H. Pope* (Guilford, Conn.: Four Quarters 1987), 219–25; citation on p. 220.

16 Josephus, *Antt.* 4.326; of Enoch, 1.85; on Mount Sinai, 3.96.

17 Philo, *Sacr. Abel. et Cain.* 8.

18 There is some question about the exact force of ἐπί here: does it mean

that humans as created were immortal, or that they were created *for* an immortality which was potential? For a more detailed discussion of these aspects see J. Barr, *Biblical Faith and Natural Theology* (Oxford: Clarendon Press 1993), pp. 61f. In Wisdom it is the righteous who gain immortality. Note, incidentally, its conviction that they only *seem* to die (Wisdom 3.2).

19 See J. Barr, *Holy Scripture: Canon, Authority, Criticism* (Oxford: Clarendon Press 1983).

20 On these two points see J. Barr, *Biblical Faith and Natural Theology* (Oxford: Clarendon Press 1993), pp. 39–80.

21 M. A.Knibb, 'Life and Death in the Old Testament', in R.E. Clements (ed.), *The World of Ancient Israel* (Cambridge University Press 1989), pp. 395–415; quotation from pp. 402f. I have gone farther than Professor Knibb, for I have argued that in Genesis itself death was *not* in fact prescribed as the penalty for disobedience.

2 *The Naturalness of Death, and the Path to Immortality*

1 For this argument, cf. C.F. Evans, *Resurrection and the New Testament* (London: SCM Press 1970), pp. 11ff.

2 G.W. Nickelsburg, *Resurrection, Immortality, and Eternal Life in Intertestamental Judaism* (Cambridge, Mass.: Harvard Theological Studies 26, 1972), p. 180.

3 In this respect it may be that Cullmann reverted to a more crude and old-fashioned dogmatic position than that of Karl Barth, with whom he otherwise has much in common. In particular, Barth, perhaps surprisingly, accepts that there is a side of death in the Bible that is natural. 'It belongs to human nature ... that man himself should be mortal' (*Church Dogmatics*, III/2, Edinburgh: T.&T.Clark 1960, p. 632). While death in the Bible 'generally' is dominated by sin, conflict, apostasy, curse and punishment, 'another form of death' is also to be seen there. When David says 'I go the way of all flesh', he 'obviously has no idea that death is a curse, or that things ought to be different' (ibid., pp. 633–4). Barth then has trouble in explaining how these two kinds of death are distinguished. The 'natural' death is a 'remarkable change of standpoint connected with a particular and extraordinary intervention on the part of God'. This 'extraordinary intervention' by God, in order to provide what is obviously perfectly normal, is required, one may suggest, not by the biblical texts, but exactly by Barth's own persistence, against the Genesis text, in believing that death was Adam's punishment for disobedience. On Cullmann's worries about Barth's admission of a natural death, see his *Immortality of the Soul*, pp. 28n., 56.

4 Gerleman, *THAT* I, 549–57, see especially cols. 554f.
5 Ibid.
6 I translate שלום here as 'goodness'; RSV 'weal'. It is interesting that the major Isaiah Scroll has טוב, which has influenced me in this direction.
7 See for example the Targum text of this passage.
8 L. Wächter, *Der Tod im Alten Testament* (Stuttgart: Calwer Verlag 1967), devotes considerable space to this aspect, 'acceptance and affirmation of death', cf. especially pp. 56–79.
9 On mourning customs, see the fine article of Eileen de Ward, 'Mourning Customs in 1, 2 Samuel', *Journal of Jewish Studies* 23, 1972, pp. 1–27 and 145–166.
10 On this see U. Kellermann, 'Überwindung des Todesgeschicks in der alttestamentlichen Frömmigkeit vor und neben dem Auferstehungsglauben', *ZThK* 73, 1976, pp. 269f.
11 My own translation from the Hebrew text, but reading the first word הבל as אבל 'mourning' with the Greek.
12 Cullmann, *Immortality*, p. 33.
13 Ibid.
14 The best thinking that I have seen along these lines comes from Otto Kaiser: see his 'Die Zukunft der Toten' and his 'Tod, Auferstehung und Unsterblichkeit' (full details in the Bibliography). Kaiser is free from the polemical hostility against Greek thought which has damaged so much twentieth-century biblical scholarship. Cf. also C.K.Barrett, 'Immortality and Resurrection', *London Quarterly and Holborn Review* 190, 1965, 91–102; he writes: 'Early Greeks and early Hebrews were markedly similar in their outlook upon physical death and what lay beyond it' (p. 92).
15 I am aware that the ability of the shades to speak may have been brought about in part by the literary necessities of Homer's poetry, and that there are other traditions according to which they could not speak but only squeaked or chirped. In the Bible this has been connected with Isaiah 8.19 and 29.4. In the Odyssey the shades of the slain suitors are led away squeaking like bats (24.5–10). See J.Bremmer, *The Early Greek Concept of the Soul* (Princeton 1983), pp. 84f. But if Homer could imagine the shades as speaking intelligibly, Hebrew writers could do the same. The shades of Isaiah 14.10 spoke excellent Hebrew in good poetical style. The case of Samuel with the witch of Endor is an interesting one, for it has a marked parallelism with that of Teiresias in the Odyssey. He alone among the dead had a νόος (*Odyssey* 10, 494–5). Bremmer, p. 84, writes: 'As a seer he fell outside the normal community of the living ... for that reason he also stands outside the community of the dead.' This might be relevant to the interesting problem why Samuel is the only one in the Hebrew Bible to do this sort of thing.

16 Kellermann, 'Überwindung', p. 277, says that in the use of this term actual *removal* is not meant, but this exclusive destiny (*Geschick*) serves as a picture to express the continuing community of the pious with Yahweh.
17 For a recent summary see Knibb, 'Life and Death in the Old Testament', pp. 403–7.
18 So for example H. Birkeland, 'The Belief in the Resurrection of the Dead in the Old Testament', *Studia Theologica* 3. 1949, pp. 69f.: 'actual death cannot have been meant ... Life expresses the positive and Death the negative aspect of existence.' But see Kapelrud's review of C. Barth's book in *SEÅ* 13, 1948, and Birkeland's judgment that Barth applied Pedersen's views in a 'somewhat mechanical' way.
19 For a recent summary cf. Knibb, 'Life and Death in the Old Testament', pp. 397ff.
20 A.R. Johnson, *The Vitality of the Individual in the Thought of Ancient Israel* (Cardiff: University of Wales, 1964), p. 84, quoted in Knibb, 'Life and Death', p. 397. Knibb writes of this: 'He [Johnson] took as the starting point of his argument the idea that Israelite thinking, like that of so-called "primitive" peoples, was predominantly synthetic.' Whether that is a good foundation may well be questioned today. Moreover, from what one suspects about 'primitive' peoples, they often had, along with a synthetic way of thinking, a complicated and mysterious set of ideas about souls and their modes of independence from the body.
21 Knibb, 'Life and Death in the Old Testament', p. 398.
22 See the idea of 'illegitimate totality transfer' in my *The Semantics of Biblical Language* (London: Oxford University Press 1961, reissued London: SCM Press 1983), pp. 218, 222. This fault in method, then identified, continues to appear. Barth's handling of *nephesh (Church Dogmatics* III/2, pp. 378f.) is an outstanding example.
23 C. Westermann, *THAT* II, 71–96.
24 See Westermann's capable article in *THAT* II, 71–96, in which he carefully separates out various realms and components of meaning, of which only one group, he says, is properly rendered by the German *Seele*.
25 Westermann, col. 94, explains this as a 'cultic-rhetorical formation, in which *nephesh* no longer has any meaning of its own but only the transformed meaning of the personal pronoun'. I understand this, but do not see why it must be so. The person can, I think, perfectly well address his soul, and this is what he does here.
26 Westermann, col. 84.
27 In some of these cases modern translations, puzzled over how to render *nephesh*, offer various renderings which make it difficult to recognize the actual Hebrew wording.
28 Proverbs 11.17 is another case, but with the less common word שאר for 'flesh'.

29 Hans Walter Wolff, *Anthropology of the Old Testament* (London: SCM Press 1974), pp. 10–25.
30 My argumentation here has parallels with that of J.F.A. Sawyer, 'Hebrew Words for the Resurrection of the Dead', *VT* 23, 1973, 218–34. What was written in one way could later be interpreted in another. His title specifies resurrection, but I think the argument could accommodate immortality of the soul equally well. The characteristic rabbinic terms like 'have a part in the world to come' are far from precision over differences of this kind.
31 Benedikt Otzen, 'Old Testament Wisdom Literature and Dualistic Thinking in Late Judaism', *VTS* 28, 1975, 146–157; citation from p. 147.
32 Josephus, *BJ*. 2,154, 163. Stendahl, in his Introduction to *Immortality and Resurrection* (New York: Macmillan 1958), pp. 7f., suggests that this may be a 'drastically apologetic translation of their belief in the resurrection', in other words that Josephus has recast in Greek terms a faith that was really in the resurrection, rather than in the immortality of the soul. I much doubt this. In any case, even if the suggestion is right, it makes no difference: for (*a*) if Josephus was recasting Jewish beliefs in this way, there is no reason in principle why New Testament writers might not be doing the same thing; (*b*) if Josephus was doing it, he had plenty of earlier and genuine Jewish precedents upon which he could have modelled these 'Greek' descriptions.
33 Cullmann, *Immortality*, p. 36, is surely utterly unconvincing when, in support of his theory, he interprets this to mean *God* as the one who can kill both soul and body. It is vastly more likely that the devil is meant. In this he seems to follow Barth, *Church Dogmatics* III/2, pp. 354, 379. In any case, one way or the other, the text differentiates between soul and body in a very serious way.
34 Westermann, THAT II, 95; N.P. Bratsiotis, '*Nepheš-psychè*, ein Beitrag zur Erforschung der Sprache und der Theologie der Septuaginta', *VTS* 15, 1966, 181–228.
35 Stendahl has been properly critical of this aspect: 'the "Greek" of this dichotomy is not the world of Homer or the Tragedians but that very special kind of Platonism and Stoicism which merged into the Hellenistic era ... or it is a catchword for Western philosophical sentiments. Furthermore, it must be noted that at no point in the history of the Christian tradition did the Church feel it to be its obligation to preserve the Semitic patterns of thought as distinct and superior to the Greek philosophical tradition' (Introduction, pp. 4–5).
36 From 'Afterlife', *Encyclopedia Judaica* 2, 336ff. [signed ED. = Scholem?], cols. 337–8.
37 Cf. Shab. 152a, which says: 'A man's soul mourns for him [after death]

for seven whole days, for it is said, *And his soul mourneth for him* (Job 14.22: ונפשו עליו תאבל).

38 See A. Kasher and S. Biderman, 'Why was Baruch de Spinoza Excommunicated?', in D.S. Katz and J.I. Israel (eds.), *Sceptics, Millenarians and Jews* (Leiden: Brill 1990), 98–141: reference on pp. 106–8.

39 B. Lifschitz, 'Inscriptions grecques de Césarée en Palestine (Caesarea Palestinae)', and 'La vie d'au delà dans les conceptions juives. Inscriptions grecques de Beth Shearim', *Revue Biblique* 68, 1961, pp. 115–26 and 401–11; quotations from pp. 117f., 405.

40 See my article '"Thou art the Cherub" (Ezekiel 28.14)', forthcoming in a Festschrift awaiting publication.

41 For a recent discussion of this portion of Ezekiel see R.R. Wilson, 'The Death of the King of Tyre: the Editorial History of Ezekiel 28', in Marks and Good, *Love and Death*, pp. 211–8. Earlier, Herbert G. May, 'The King in the Garden of Eden: a Study of Ezekiel 28:12–19', in B.W. Anderson and W. Harrelson (eds.), *Israel's Prophetic Heritage: Essays in Honor of James Muilenburg* (New York: Harper and London: SCM Press 1962), 166–76.

42 On this see Kellermann, 'Überwindung', p. 282.

43 On all this section see already my article 'The Authority of Scripture: the Book of Genesis and the Origin of Evil in Jewish and Christian Tradition', in *Christian Authority* (Henry Chadwick Festschrift, Oxford: Clarendon Press 1988), pp. 59–75.

44 Since many readers of the Bible remain unaware of the importance of this material, they may think that my emphasis on it is something novel and idiosyncratic. To them I would point out that so traditional and respectable a work as N.P. Williams, *The Ideas of the Fall and of Original Sin* (London: Longmans, 1927), gives considerable space and attention to the 'Watcher-theory' and writes (p. 27) that it was 'the earliest and crudest form in which the doctrines of the Fall and of Original Sin obtained any wide currency among the Jews'.

45 Nickelsburg, *Resurrection, Immortality and Eternal Life*, p. 179. He comments: 'There is no hint that these must or will take on bodies in order to experience fully the eternal life.'

46 Stendahl, *Meanings*, Introduction, p. 194.

47 I am gratified to find a similar opinion in Kellermann, 'Überwindung', if I understand him rightly. See his concluding section, pp. 281–2.

3 *Knowledge, Sexuality and Immortality*

1 The word order is important here, for it is a clue to a possible combination of two elements. This is obscured by harmonizing renderings such as that of NIV with its 'In the middle of the garden were the tree of life and

the tree of the knowledge of good and evil.'

2 The contrary view has been held by one or two scholars; but see the classic survey by Paul Humbert, *Études sur le récit du paradis et de la chute dans la Genèse* (Neuchâtel 1940), pp. 131f. Like Humbert, I have gone through all the 131 cases of Hebrew פן 'lest' in the Bible and found none which means 'lest someone continues to do what they are already doing'. Moreover, the expression 'put out his hand and do something' is an inchoative expression and cannot easily mean 'continue to do what he has been doing all along'.

3 For a recent discussion cf. C. Westermann, *Genesis 1–11* (Minneapolis: Augsburg Publishing House and London: SPCK 1984), pp. 213f.

4 See the survey in Westermann, *Genesis 1–11*, pp. 242–5.

5 Already briefly mentioned in ch. 1, p. 11.

6 Westermann, *Genesis 1–11*, p. 270.

7 See James Barr, 'The Symbolism of Names in the Old Testament', in *Bulletin of the John Rylands Library* 52, 1969–70, 11–30.

8 Cf. Westermann, *Genesis 1–11* pp. 268f. This suggestion has had the good fortune that it was already thought of and experimented with by rabbis in ancient times. If this had not been so, the modern scholars who embraced it, like Wellhausen, would have been excoriated as 'historicists' for thinking of it.

9 Westermann, *Genesis 1–11*, p. 243.

10 Six hours were all that was accorded to the time in the Garden of Eden by many traditions: cf. J.G. Turner, *One Flesh: Paradisal Marriage and Sexual Relations in the Age of Milton* (Oxford: Clarendon Press 1987), p. 19. St Augustine held that the disobedience followed 'as soon as the woman was created', so that there was no time for making love: Turner, p. 29. Milton had 'the first couple live for weeks in Paradise enjoying full sexual intercourse without pregnancy', ibid., p. 30.

11 Jubilees indeed (3.34) follows the Genesis text in saying that 'They had no son till the first jubilee; and after this he had intercourse with her'.

12 Harold Bloom, *The Book of J* (New York: Weidenfeld 1990), p. 183. I am not sure, however, that this interpretation has been as 'normative' as Bloom supposes.

13 Gen. R., xxii.2; *Midrash Rabba* (London: Soncino 1977), p. 180.

14 See J. Barr, 'Philo of Byblos and his "Phoenician History"', *Bulletin of the John Rylands Library* 57, 1974–5, pp. 17–68.

15 On this see Westermann, *Genesis 1–11*, p. 269.

16 See Gunnlaugur Jónsson, *The Image of God. Genesis 1:26–28 in a Century of Old Testament Research* (Coniectanea Biblica, Old Testament Series 26, Lund: Almquist & Wiksell 1988); Jónsson is so kind as to say (p. 146 n. 6) that 'if Karl Barth is the most important name for *imago Dei* studies during the period 1919–1960, it may with some

justice be claimed that James Barr is the most important name during the period 1961–1982'. It is not easy to live up to such an evaluation!

4 *Noah's Ark: Time, Chronology and the Fall*

1 Stephanie Dalley, *Myths from Mesopotamia* (Oxford University Press 1989), p. 118.
2 This is correctly reported and discussed by Lloyd R. Bailey, *Noah: the Person and the Story in History and Tradition* (Columbia: University of South Carolina 1989), pp. 165ff. The objection, voiced by him (p. 166), that it would imply a new verb or a meaning which has survived only in Ethiopic can be answered by suggesting that 'rest' and 'be long (of time)' can be plausibly considered as relatable components of meaning. Cf. for instance the sense 'to be slow, still' recorded 'in the stative' of *nâḫu* in the *Chicago Assyrian Dictionary*, vol. N. 1, 1980, p. 144, with the examples *šumma ina alākišu ne-eḫ* 'if he walks slowly' and *šumma Sin ina alākišu ni-iḫ* 'if the moon is slow in its course'. One would not, therefore, have to hypothesize the existence of a completely different verb.
3 'The name נח has not been explained', says Westermann, *Genesis 1–11*, p. 360. He appears not to know of the explanation I follow here.
4 It should be noticed that the term I have translated as 'formation', Hebrew *yeṣer*, is the same word used for the two contrary 'inclinations' or 'impulses' in post-biblical Judaism, mentioned in Ch. 2 above.
5 On this theme cf. J.D. Levenson, *Creation and the Persistence of Evil* (San Francisco: Harper and Row 1988).
6 J. Bremmer, *The Early Greek Concept of the Soul* (Princeton 1983), pp. 120f.
7 Deucalion was, of course, the Noah of the Greek Flood.
8 H.F.D. Sparks, *The Apocryphal Old Testament*, pp. 190ff. (translated by M.A. Knibb).
9 This is relevant also to the *extension* of the severe words about human evil I have quoted from Genesis 6.5. These words have a *likeness* to later ideas of a *general* human depravity, but were not necessarily meant to be general in that sense. It was meant of the generation of the flood that all their thoughts were 'only evil all the time', and this did not mean that it applied to all human beings of all time past and future. Even at that time it did not apply to Noah, who is expressly described as 'righteous' and even 'perfect' or at least 'blameless', תמים, and this righteousness extended to his family. On the other hand God's words at Genesis 8.21 seem to accept that inveterate human evil will continue into the future, and that it will be of no use to try to wipe it out by another flood or similar drastic action.

10 On this see my articles 'Why the World was Created in 4004 BC: Archbishop Ussher and Biblical Chronology', and 'Luther and Biblical Chronology', *Bulletin of the John Rylands Library* 67, 1984–85, 575–608, and 72, 1990, 51–67, and *Biblical Chronology: Science or Legend?* (Ethel M. Wood Lecture, London: Athlone Press, 1987). Going back earlier, see J. Barr, *Biblical Words for Time* (London: SCM Press 1969).

11 See the fine article by B.S. Childs, 'Adam', in *Interpreter's Dictionary of the Bible* I, p. 43.

12 See H.-J. Birkner, 'Natürliche Theologie und Offenbarungstheologie: ein theologiegeschichtlicher Überblick', *Neue Zeitschrift für Systematische Theologie* 3, 1961, 279–95; reference on p. 282.

13 See J. Barr, 'The Authority of Scripture: The Book of Genesis and the Origin of Evil in Jewish and Christian Tradition', in *Christian Authority* (Henry Chadwick Festschrift, Oxford: Clarendon Press 1988), pp. 59–75, reference on p. 74.

14 Calvin, *Genesis* (Edinburgh: Calvin Translation Society 1847), 238. His comment is: 'That ancient figment, concerning the intercourse of angels with women, is abundantly refuted by its own absurdity, and it is surprising that learned men should formerly have been fascinated by ravings so gross and prodigious.' For Luther, see my article 'Luther and Biblical Chronology', *Bulletin of the John Rylands Library* 72, 1990, 51–67.

15 Cullmann, *Immortality* p. 21. The contrast between the deaths of Socrates and Jesus was a stock theme in much of the inheritance of the dialectical theology. For a contrasting treatment of Socrates' death by a biblical scholar, see Otto Kaiser, 'Der Tod des Socrates', in his *Der Mensch unter dem Schicksal* (Berlin: de Gruyter 1985), 196–205.

16 Cullmann, *Immortality*, pp. 21–4.

17 See Henning Graf Reventlow, *Problems of Old Testament Theology in the Twentieth Century* (London: SCM Press and Philadelphia: Fortress Press 1985), pp. 23ff. The essential arguments of Köhler, as summarized below, are in the 8 July 1926 issue of *KBRS*, pp. 105f.

18 *KBRS*, 22 July 1926, pp. 113f.

19 Cf. Christof Gestrich, *Neuzeitliches Denken und die Spaltung der dialektischen Theologie* (Tübingen: Mohr 1977), p. 35, writes: 'But Brunner was far from open to that which is called the "historical character" of truth. He had a strong tendency to think in timeless structures of essence and to read timeless "forms of thought" out of the material of theology and the history of ideas. In this respect, as the most energetically "anti-historical" representative of dialectical theology, he really reminds us of a rationalistic or medieval thinker, little bothered by the historical.'

20 W. Sibley Towner, 'Interpretations and Reinterpretations of the Fall', in F.A. Eigo, *Modern Biblical Scholarship: Its Impact on Theology and Proclamation* (Proceedings of the Theology Institute of Villanova University 1984), p. 81.
21 Gene M. Tucker, 'The Creation and the Fall: a Reconsideration', *Lexington Theological Quarterly* 13, 1978, 113–24; citation from p. 122.
22 Towner, 'Interpretations', p. 82.

5 *Immortality and Resurrection: Conflict or Complementarity?*

1 Stendahl, *Meanings*, p. 199. These words echo the title of his article: 'Immortality is Too Much and Too Little'.
2 lbid., p. 197.
3 A. von Harnack, *What is Christianity*? (New York: Harper 1957), pp. 160–3.
4 Or, indeed, 'of the flesh', which upsets Cullmann badly, see *Immortality*, p. 46.
5 Stendahl, *Meanings*, p. 198.
6 Westminster Confession, ch. 32. For a balanced discussion see George S. Hendry, *The Westminster Confession for Today* (Richmond: John Knox Press 1960). In Hendry's presentation the same chapter is ch. 34.
7 It is characteristic that the Confession, in spite of the diligence it normally shows in assembling scriptural 'proofs' of doctrine, offers no such proofs of the immortal subsistence of the soul. For the sentence quoted, it offers only two passages: Luke 23.43, where Jesus tells the thief that 'today you will be with me in paradise', and Eccl. 12.7, which tells that 'the spirit returns to God who gave it'. The Lucan passage certainly shows the immediate translation of the dying man into the presence of God. But it does not prove the 'immortal subsistence' of the soul. That is at best a deduction, and not a statement of scripture. It does not prove that all human souls have this immortal subsistence, for God might have performed an exceptional miracle in the case of this thief. The quotation from Ecclesiastes makes clear that the spirit returns to God, but does not show that this constitutes immortal subsistence: it could mean only that the spirit is taken back into the being of God and then ceases to have independent existence. What the Confession was really interested in, of course, was the *immediate* passage of humans to their eternal destiny, in other words, in the removal of all thoughts of purgatorial existence and the like. Neither of the passages quoted, even under the conditions of the time, demonstrated 'immortal subsistence'.

As I say, this was regarded as a manifestly essential foundation of religion, which required no demonstration or proof, even from scripture.

8 S. Tugwell, *Human Immortality and the Redemption of Death* (London: Darton, Longman and Todd 1990), p. 149.
9 Hendry, *Westminster Confession*, pp. 243ff.
10 Hendry, *Westminster Confession*, pp. 245f.
11 W.S. Bouwsma, *John Calvin* (New York: Oxford University Press 1988), p. 80.
12 Ibid.; the reference given in n. 125 is to the Commentary on John 5.28.
13 Hendry, *Westminster Confession*, p. 249.
14 John Baillie, *And the Life Everlasting* (London: Faber and Faber and New York: Scribners 1933), p. 299.
15 In *Old and New in Interpretation* (London: SCM Press 1966, [2]1982), p. 54 n. 1. See also my more recent discussion in *Biblical Faith and Natural Theology* (Oxford: Clarendon Press 1993), ch. 2.
16 On the place of natural theology in this, see my *Biblical Faith and Natural Theology*, and earlier my Mowinckel Lecture, *Studia Theologica*.
17 See R. Meyer, *Hellenistisches in der rabbinischen Anthropologie. Rabbinische Vorstellungen vom Werden des Menschen*, BWANT 76 (Stuttgart 1976).
18 Josephus, *BJ* 2.163, 3.374, *c. Apion.* 2.218.
19 The relevance of this is being increasingly seen as newspaper articles report 'finds' from the Dead Sea Scrolls which allegedly indicate that some legendary figure had done, or was expected to do, things closely corresponding to what Jesus is supposed to have done. Such reports are relished for the trouble they are supposed to give to Christianity.
20 Pierre Benoît, in P. Benoît and R. Murphy, *Immortality and Resurrection* (New York: Herder and Herder 1970), p. 112.
21 Cf. already above, pp. 44f. Cf. Cullmann, *Immortality*, p. 36. Cullmann seems to be confused at this point: he insists that the saying of Jesus 'in no way presupposes the Greek conception' but has to go on to admit the distinction of soul and body and to say that 'the soul is the starting-point of the resurrection ... since ... it can already be possessed by the Holy Spirit in a way quite different from the body'.
22 Stendahl, *Meanings*, p. 196.
23 Ibid.
24 Cullmann, *Immortality*, p. 46.
25 So already above, pp. 53f. Cf. Stendahl, *Meanings*, p. 197.
26 Also in my *Old and New in Interpretation*, pp. 52ff.
27 Maurice Wiles, *The Remaking of Christian Doctrine* (London: SCM Press 1974), p. 129.

28 See Westermann, *Genesis 1–11*, pp. 260–1, for a crushing rebuttal of all such suggestions. It is interesting to note that he mentions how, among older exegetes, Calvin was the most reserved towards the interpretation of this text as christological.

29 H.C.C. Cavallin, *Life after Death. Paul's Argument for the Resurrection of the Dead in I Cor 15* (Coniectanea Biblica, New Testament Series 7:1, Lund: Gleerup 1974), p. 81.

Index of Biblical References

APOCRYPHA AND PSEUDEPIGRAPHA

Index of Names and Subjects